Current Topics in Microbiology 233/II and Immunology

Editors

Springer
Berlin
Heidelberg
New York
Barcelona
Budapest
Hong Kong
London
Milan
Paris
Santa Clara
Singapore
Tokyo

Reoviruses II

Cytopathogenicity and Pathogenesis

Edited by K.L. Tyler and M.B.A. Oldstone

With 45 Figures

Springer

Kenneth L. Tyler M.D.

University of Colorado
Health Sciences Center
Department of Neurology
Campus Box B-182
4200 East 9th Avenue
Denver, CO 80262
USA

Michael B.A. Oldstone, M.D.

The Scripps Research Institute
Department of Neuropharmacology
Division of Virology
10550 N. Torrey Pines Road
La Jolla, CA 92037
USA

Cover Illustration: Surface-shaded views of reovirus T1L showing (top) a "split" particle composed of a T1L virion and an ISVP and (bottom) a "split" particle composed of a T1L ISVP and a core particle. Pictures are based on three-dimensional image reconstruction of data obtained by cryoelectron microscopy (See Figure 3, Chapter 1 for additional details. Images provided by Dr. M.L. Nibert).

Cover Design: design & production GmbH, Heidelberg

ISSN 0070-217X
ISBN-13:978-3-642-72097-0 e-ISBN-13:978-3-642-72095-6
DOI: 10.1007/978-3-642-72095-6

Softcover reprint of the hardcover 1st edition 1998

Library of Congress Catalog Card Number 15-12910

Typesetting: Scientific Publishing Services (P) Ltd, Madras

SPIN: 10643664 27/3020 – 5 4 3 2 1 0 – Printed on acid-free paper

Foreword

Mammalian reoviruses do not cause significant human disease, yet they have commanded considerable interest among virologists, geneticists, and biochemists. Their genome consists of ten segments of double-stranded RNA and, because a number of reovirus strains exist, it has been possible to isolate and study the function and structure of a single gene of one strain placed upon the backbone of the nine other genes of the second strain using genetic reassortant techniques. Such manipulations have led to fundamental studies on binding, entry, replication, transcription, assembly, and release. In addition, reovirus type 1, compared to type 3, displays different disease phenotypes in the mouse. Reassorting genes between reovirus 1 and 3 allows a dissection of several important questions concerning host–virus interactions and understanding the molecular basis of the associated disease. Most of the major players who have made seminal contributions in this area have contributed chapters to these two volumes of Current Topics in Microbiology and Immunology 233. One whose presence is found throughout both volumes, but is not an active contributor, is the late Bernard Fields, who died on 31 January 1995.

Bernie Fields had a profound influence not only in the reovirus field, but in the arena of virology as a whole, especially viral pathogenesis. It was his appreciation of colleagues in the reovirus field and their appreciation of him, coupled with his strong commitment to the training of independent scientists, that made these two volumes dedicated to him a labor of love. Fields' significant research in this area is matched by the accomplishments of a number of committed and productive biomedical scientists that he trained, many of whom have contributed to these two volumes. It is these scientific children who will keep Fields' memory alive. Specific thanks is given to Ken Tyler, a former student and associate of Bernard Fields who was primarily responsible for organizing and collecting papers for these two volumes.

Not only do these two volumes stand as a testimony to the respect Bernard Fields earned and received from others, but the authors have unanimously agreed that their honorarium received from Springer-Verlag for the publication of these two volumes will be used to provide lecture series on microbial (viral) pathogenesis to be given in honor of Bernie.

La Jolla, California
1998

MICHAEL B.A. OLDSTONE

Preface

There were two common and interlocking themes that guided the selection of contributors to these two volumes, namely, Dr. Bernard N. Fields and reoviruses. Dr. Fields' untimely death at the age of 56 on 31 January 1995 was a tremendous loss to his family, to his friends and colleagues, to the field of virology in particular, and to science in general. All of the contributors to these volumes had close ties to Bernie, and all of us still mourn his loss. I suspect that they, as I, frequently still find themselves starting to pick up the telephone to tell Bernie about an exciting new finding or to get his advice about a perplexing scientific, professional, or personal problem. Bernie was always available for these phone calls, and was genuinely proud and excited about what his former students and trainees and his colleagues were accomplishing.

Many of the contributors to these two volumes trained as postdoctoral fellows in Dr. Fields' laboratory, including Drs. Nibert, Coombs, Ramig, Schiff, Brown, Dermody, Tyler, Sherry, Rubin, and Virgin (listed in the order in which their chapters appear). Others were collaborators, close colleagues, and friends (Drs. Shatkin, Joklik, Lee, Greene, Jacobs, Maratos-Flier, and Samuel). In many cases these individuals have enlisted colleagues and trainees from their own laboratories as coauthors. Bernie was extremely proud of the continuing multigenerational expansion in the reovirus "family," and the extraordinarily high quality of work that was being performed by so many talented researchers.

Difficult choices must inevitably be made in selecting contributors to a work of this type, and as a result inadvertent omissions occur. For these the editors take full responsibility, and offer apology in advance to any who may feel slighted. It is undoubtedly a further tribute to Bernie that everyone invited to contribute to these volumes accepted immediately and enthusiastically. It is sad, though (and the bane of all editors) that not all those who accepted were ultimately able to contribute. The editors and contributors have agreed to forego any royalties in conjunction with these volumes and instead to utilize these funds to endow a Fields' Lectureship in Microbial Pathogenesis to be held in conjunction with the FASEB meetings.

Perhaps the hardest editorial decision, required for reasons of space and logical coherence, was to restrict contributors to those still actively involved in reovirus research. Many of Bernie's former trainees who were excluded in this way have made eminent contributions in other branches of virology, immunology, and molecular biology or in other fields of science. I can only hope that their voices will be heard in future volumes dedicated to Bernie's memory.

Reoviruses have been and continue to be an important viral system for understanding the molecular and genetic basis of viral pathogenesis, a theme that was central to much of Bernie's own laboratory research. Bernie clearly recognized that understanding basic aspects of the structure, molecular biology, and replication strategy of viruses is critical to developing a complete and accurate picture of pathogenesis. From a personal perspective, I always remember Bernie's infectious excitement when he thought that some new research finding or observation helped link some fundamental aspect of basic virology with an improved understanding of how viruses ultimately cause disease. In this spirit, the contributions to these volumes run the gamut from studies in basic reovirology to the use of reoviruses to explore pathogenesis in vivo.

Chapters in Volume I deal with fundamental reovirology, including studies of virion structure, the structure and function of individual viral structural proteins, and the nature of the virus genome and its assembly. Also included in the first volume are chapters dealing with temperature-sensitive (ts) mutants, and the effects of reovirus interaction with cell surface receptors.

In Volume II the focus shifts to emphasize the effects of reoviruses' interaction with target cells and of reovirus infection on individual organ systems. This second volume includes chapters dealing with the molecular mechanisms of reovirus persistent infections and reovirus-induced apoptosis. These chapters are followed by individual chapters dealing with reovirus infection of particular organs including the heart, liver, biliary, endocrine, and nervous systems. These selections are not intended as an exhaustive catalogue of pathology but rather to highlight reoviruses' diverse effects on many biological systems in vivo and the mechanisms by which these occur. Finally, as Bernie clearly recognized, it is impossible to truly understand viral pathogenesis without also understanding the role played during infection of the various components of the host's immune system. The concluding chapters of the second volume deal with selected aspects of reoviruses and their interaction with cytokines, antibodies, and the cellular immune system.

These two volumes should be considered a selective snapshot of the current state of the art in many areas of reovirus research. All such collections suffer from innate biases, with some topics being overemphasized and others inadvertently omitted. For all these imperfections the editors take full responsibility. We can only hope that the readers of these volumes will find that the pleasure obtained from seeing so much good work being carried out by so many gifted people will outweigh the annoyances engendered by any deficiencies. We can also only hope that Bernie, were he still alive, would have felt the same way and considered this a sort of *Festschrift*, celebrating the profound influence he had on all of us.

Denver, Colorado
1998

KENNETH L. TYLER

List of Contents

List of Contents of Companion Volume 233/I

List of Contributors

(Their addresses can be found at the beginning of their respective chapters.)

Molecular Mechanisms of Persistent Infection by Reovirus

T.S. DERMODY

1 Introduction

Many cytolytic animal viruses are capable of establishing persistent infections of cultured cells. For such persistent infections to be maintained interactions between virus and cell must be modulated such that a less cytopathic virus-host relationship is established (reviewed in AHMED et al. 1996). Variant viruses may be selected that are attenuated in cytolytic potential, or variant cells may be selected that are less permissive for viral replication. In some cases, however, viruses and cells coevolve during persistent infection, such that selection of virus-resistant cells leads to counterselection of highly infective mutant viruses that can grow in resistant cells.

Studies of these coevolving cultures of viruses and cells have identified key steps in virus-cell interaction that are modified by cells to resist cytolytic infection. Specifically, steps in viral replication required for viral entry are targeted in persistent infections caused by several viruses, including coronavirus (GALLAGHER et al. 1991;

Departments of Pediatrics, and Microbiology and Immunology, and Elizabeth B. Lamb Center for Pediatric Research, Vanderbilt University School of Medicine, Nashville, TN 37232, USA
(E-mail: terry.dermody@mcmail.vanderbilt.edu)

Chen and Baric 1996), poliovirus (Kaplan et al. 1989; Borzakian et al. 1992), reovirus (Dermody et al. 1993; Wetzel et al. 1997a), and rotavirus (Mrukowicz et al., 1998). Identification of the specific strategies used by viruses to overcome cellular blocks to viral entry has provided new insights into mechanisms by which viruses enter cells.

2 Persistent Reovirus Infections are Carrier Cultures That Require Horizontal Cell-to-Cell Transmission for Their Propagation

Although usually cytolytic in cell culture, mammalian reoviruses can establish persistent infections of many types of cells, including human embryonic fibroblast cells (Bell et al. 1966), Burkitt's lymphoma cells (Levy et al. 1968), Chinese hamster ovary (CHO) cells (Taber et al. 1976), murine L929 (L) cells (Ahmed and Graham 1977; Ahmed and Fields 1982; Brown et al. 1983; Dermody et al. 1993), murine B- and T-cell hybridomas (Matsuzaki et al. 1986; Dermody et al. 1995), murine 3T3 cells (Verdin et al. 1986), Madin-Darby canine kidney (MDCK) cells (Montgomery et al. 1991), murine SC1 cells (Danis et al. 1993), and murine erythroleukemia (MEL) cells (Wetzel et al. 1997a). Cell cultures persistently infected with reovirus produce high titers of virus for long periods of time, and a majority of cells in the cultures shows evidence of viral infection (Taber et al. 1976; Ahmed and Graham 1977; Verdin et al. 1986; Dermody et al. 1993; Wetzel et al. 1997a). Anti-reovirus antibody treatment of persistent infections of CHO cells (Taber et al. 1976), L cells (Ahmed et al. 1981; Dermody et al. 1993), B-cell hybridomas (Dermody et al. 1995), and MEL cells (Wetzel et al. 1997a) results in cure of persistent infection.

The observation that antibody treatment can cure cell cultures persistently infected with reovirus suggests that persistent reovirus infections are maintained by horizontal transmission of virus between cells (Mahy 1985). Antibodies are believed to neutralize viral infectivity by blocking early steps in viral replication, such as attachment, penetration, and disassembly (reviewed in Whitton and Oldstone 1996). It is likely that neutralizing anti-reovirus antibodies block one or more of these early steps, which would interrupt horizontal viral transmission and result in cure of persistent infection. It is possible that some component of vertical viral transmission occurs during persistent reovirus infection of cultured cells. However, the rapid decrease in viral titer during antibody treatment (Dermody et al. 1993, 1995; Wetzel et al. 1997a) suggests that horizontal transmission is the primary mechanism of viral spread in these cultures.

3 Distinct Phases of Persistent Reovirus Infection Have Been Identified

Persistent reovirus infections of L cells have been used as a useful model system to define mechanisms that foster long-term propagation of persistent viral infections.

From studies of persistently infected L-cell cultures, a general paradigm has emerged which holds that there are two distinct phases of persistent infection: establishment and maintenance. Establishment of persistent infection is characterized by intense cell crises in which only a few colonies of cells survive. Maintenance is characterized by stable cell growth and continuous production of substantial titers of infectious virus (ca. 1×10^6–1×10^8 plaque-forming units per milliliter of culture supernatant) for prolonged periods of passage, in some cases in excess of 3 years (T.S. DERMODY, unpublished observations).

Establishment of persistent reovirus infections of L cells occurs when infection is initiated with virus stocks passaged serially at high multiplicity of infection (AHMED and GRAHAM 1977; AHMED and FIELDS 1982; BROWN et al. 1983; DERMODY et al. 1993). Such stocks contain a variety of viral mutants (AHMED et al. 1980, 1983), and some of these mutants may facilitate establishment of persistent infection. In an early study of persistent reovirus infection, mutations in the viral S4 gene segment, which encodes outer-capsid protein σ3, were suggested to be important for establishment of persistent infection (AHMED and FIELDS 1982). In this study L cells were coinfected with a low-passage stock of reovirus strain type 2 Jones (T2J), which causes lytic infections, and a high-passage stock of strain type 3 Dearing (T3D), which causes persistent infections, under conditions to promote persistent infection. It was reasoned that gene segments selected from the T3D high-passage stock during the coinfection would identify viral mutations required for establishment of persistent infection. In three independent coinfections, the S4 gene segment was the only gene found to be consistently selected from the T3D high-passage stock (AHMED and FIELDS 1982), arguing that mutations in S4 are required for establishment of persistent infection. However, it is also possible that the T3D S4 gene confers a selective advantage over the T2J S4 gene in mixed infections of L cells, rather than mediating establishment of persistent infection.

Not all reovirus strains generate mutations capable of establishing persistent infection during high passage. Stocks of T3D, but not strain type 1 Lang (T1L), passaged at high multiplicity of infection contain deletions and other mutations and readily establish persistent infections (BROWN et al. 1983). Reassortant viruses containing a T3D L2 gene segment mediate these properties (BROWN et al. 1983), suggesting that L2 gene product λ2 is important for the generation of mutations that allow persistent infection to be established. The λ2 protein is a major component of the reovirus core (RALPH et al. 1980; DRYDEN et al. 1993), and λ2 serves as the viral guanylyltransferase (CLEVELAND et al. 1986). Therefore it is possible that λ2-mediated functions in viral RNA synthesis are the basis for differences in the mutation frequency exhibited by T1L and T3D (BROWN et al. 1983).

Other data indicate that the type of host cell determines whether persistent infection is established. Persistent infections of 3T3 cells (VERDIN et al. 1986), CHO cells (TABER et al. 1976), MDCK cells (MONTGOMERY et al. 1991), SC1 cells (DANIS et al. 1993), and MEL cells (WETZEL et al. 1997a) can be established using low-passage reovirus stocks, which lack the capacity to establish persistent infections of L cells (AHMED and FIELDS 1982; WETZEL et al. 1997a). In some cases the capacity of cells to support establishment of persistent infection is linked to resistance to reovirus-induced inhibition of cellular protein synthesis (DUNCAN et al. 1978; DANIS et al. 1993). In other cases establishment of persistent infection is favored by the inability of host cells to efficiently support acid-dependent proteolytic disassembly of reovirus virions during viral entry. Treatment of L cells with the weak base ammonium chloride, which blocks acid-dependent proteolysis of reovirus virions (STURZENBECKER et al. 1987), leads to establishment of persistent infections using low-passage reovirus stocks (CANNING and FIELDS 1983). The capacity of MEL cells to promote establishment of persistent infection is also linked to an inability to efficiently support viral disassembly (WETZEL et al. 1997a). These observations suggest that blocks to reovirus entry, either pharmacological or endogenous, favor persistent over lytic infections of cultured cells.

4 Mutations in Cells and Viruses Affecting Viral Entry Coevolve During the Maintenance Phase of Persistent Reovirus Infection

Coevolution of viruses and cells during persistent reovirus infection has been documented in studies of persistent infections of L cells (AHMED et al. 1981; DERMODY et al. 1993) and MEL cells (WETZEL et al. 1997a). Viruses selected during maintenance of these persistent infections (termed PI viruses) grow better than wild-type (wt) viruses in cells cured of persistent infection (Fig. 1), indicating that mutant viruses are selected during persistent infection. Similarly, wt viruses grow better in parental cells than in cured cells, indicating that mutant cells are also selected in these cultures (Fig. 1). Insight into the nature of mutations selected in cells and viruses during persistent reovirus infection was first suggested by electron micrographic analysis of persistently infected and cured L cells (AHMED et al. 1981; SHARPE and FIELDS 1983). Both types of cells accumulate large numbers of vacuoles that resemble lysosomes (Fig. 2). These findings led to the hypothesis that mutations in cells affect steps in reovirus entry dependent on proteolysis of the viral outer capsid in vacuoles of the endocytic compartment.

Reovirus entry is initiated by stable attachment of the virus to the surface of the host cell. The σ1 protein, which is encoded by the S1 gene segment, serves as the reovirus attachment protein (WEINER et al. 1980; LEE et al. 1981). The σ1 protein is a fibrous protein (FURLONG et al. 1988; FRASER et al. 1990) located at the 12 vertices of the virion icosahedron (FURLONG et al. 1988; DRYDEN et al. 1993). Native σ1 protein forms an oligomer, and current data suggest that the oligomeric species of σ1 is either a trimer (LEONE et al. 1991a, 1992; STRONG et al. 1991) or a tetramer (BASSEL-DUBY et al. 1987; FRASER et al. 1990). Following viral attachment, virions are observed by electron microscopy in clathrin-coated pits, which suggests that virion uptake occurs by receptor-mediated endocytosis (BORSA et al. 1979, 1981; STURZENBECKER et al. 1987; RUBIN et al. 1992). Within late endosomes or lysosomes, viral outer-capsid proteins σ3 and μ1/μ1C are subject to proteolysis by cellular proteases, resulting in generation of infectious subvirion particles

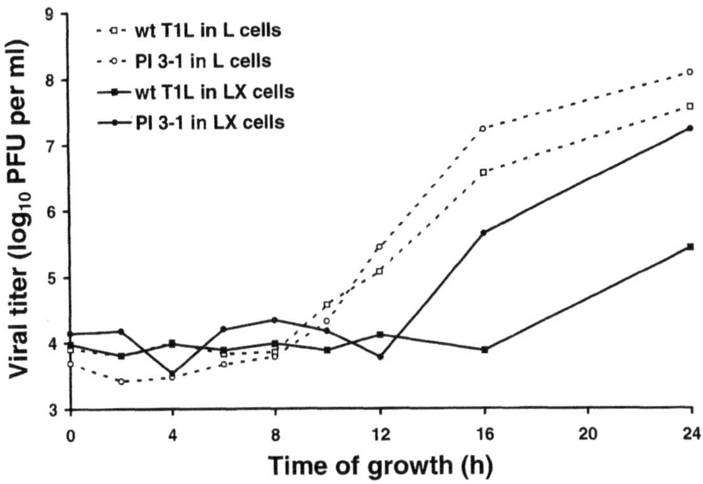

Fig. 1. Growth of wt virus T1L and PI virus PI 3-1 in parental L cells and cured LX cells. Monolayers of cells (5×10^5 cells) were infected with either T1L or PI 3-1 at an MOI of 2 PFU per cell. After a 1-h adsorption period the inoculum was removed, fresh medium was added, and the cells were incubated at 37°C for the time, shown. Cells were frozen and thawed twice, and virus in cell lysates was titrated on L-cell monolayers by plaque assay. The results are presented as the mean viral titers for two independent experiments. (Adapted from Wilson et al. 1996)

(ISVPs) (Chang and Zweerink 1971; Silverstein et al. 1972; Borsa et al. 1981; Sturzenbecker et al. 1987). During this process σ3 is degraded and lost from virions, viral attachment protein σ1 undergoes a conformational change, and μ1/μ1C is cleaved to form particle-associated fragments μ1δ/δ and ϕ (reviewed in Nibert et al. 1996). ISVPs generated in the endocytic compartment are probably identical to those generated either in the intestinal lumen of perorally infected mice (Bodkin et al. 1989; Bass et al. 1990) or in vitro by treatment of virions with chymotrypsin or trypsin (Chang and Zweerink 1971; Silverstein et al. 1972; Borsa et al. 1981; Sturzenbecker et al. 1987; Nibert et al. 1995). Intracellular proteolysis of σ3 and μ1 is an acid-dependent process as treatment of cells with the weak base ammonium chloride (Sturzenbecker et al. 1987; Dermody et al. 1993) or inhibitors of the vacuolar proton ATPase, such as bafilomycin or concanamycin A (Martinez et al. 1996), blocks infection by virions but not by ISVPs.

The availability of in vitro generated reovirus disassembly intermediates has facilitated conclusive demonstration that mutant cells are altered in their capacity to support viral entry. Cured cells do not support efficient growth of wt virus when infection is initiated with virions, but do so when infection is initiated with in vitro generated ISVPs (Dermody et al. 1993) (Fig. 3). These findings indicate that mutant cells selected during persistent reovirus infection do not support steps in viral entry leading to generation of ISVPs. The mechanisms by which mutant cells block steps in reovirus entry are not known; however, it is possible that mutant cells are altered in their capacity to bind virions, internalize virions by receptor-mediated endocytosis, mediate acid-dependent disassembly of the viral outer capsid, or

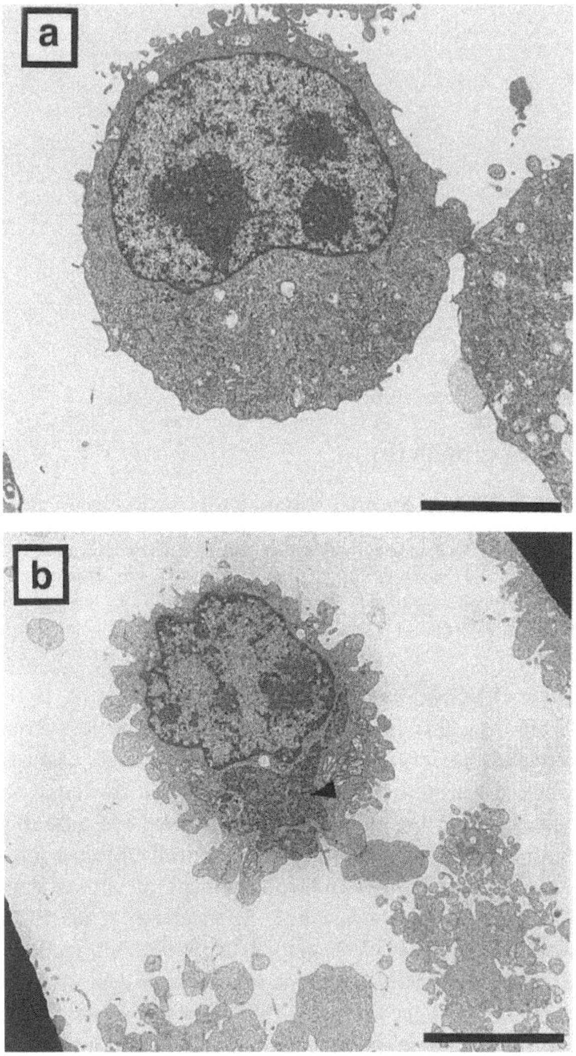

Fig. 2a–d. Morphology of uninfected L cells, persistently infected L cells, and cured LX cells. **a** Uninfected L cells. **b** Persistently infected L cells showing large inclusion of progeny virions (*arrowhead*). **c** Persistently infected L cells showing membrane-bound organelles (*arrows*). **d** Cured LX cells showing membrane-bound organelles (*arrows*). *Bars*, 5μm

facilitate penetration of the viral core into the cytoplasm. Cells manifesting alterations in endocytic function analogous to those selected during persistent reovirus infection are also observed after selection for resistance to diphtheria toxin. Organelles in the central vacuolar system of diphtheria-toxin-resistant CHO cells are altered in their acidification capacity (Merion et al. 1983). Therefore changes in endocytic uptake and proteolytic processing might be common mechanisms for

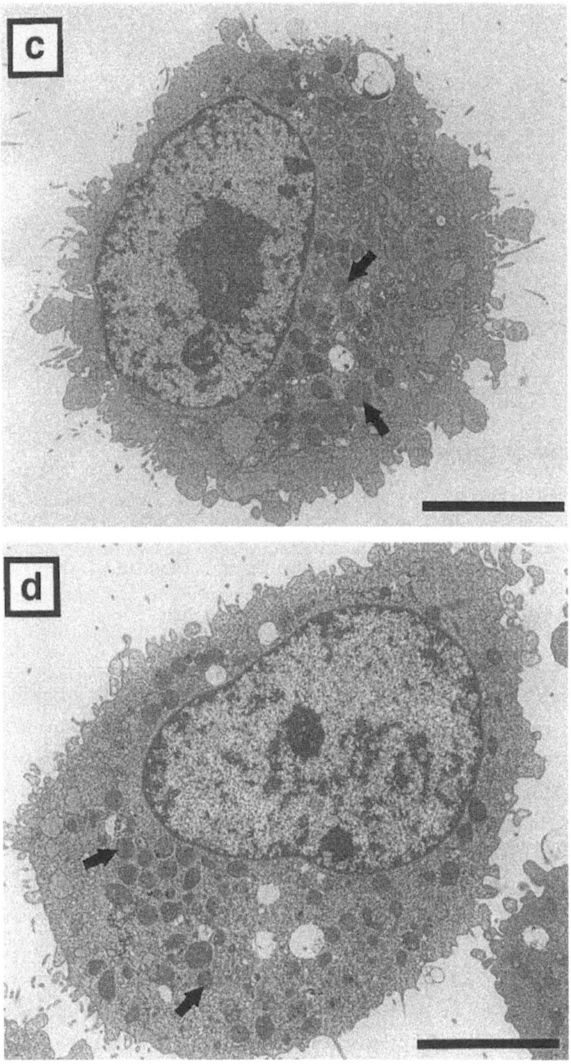

Fig. 2c–d.

cells to acquire resistance to cytotoxic substances, such as pathogenic micro-organisms or their toxins.

The finding that mutant cells selected during persistent reovirus infection do not support proteolytic disassembly of viral outer-capsid proteins led to the suggestion that mutant viruses are altered in their requirement for acid-dependent proteolysis to facilitate entry (DERMODY et al. 1993). This is indeed the case. In contrast to wt viruses, PI viruses grow well in L cells treated with ammonium

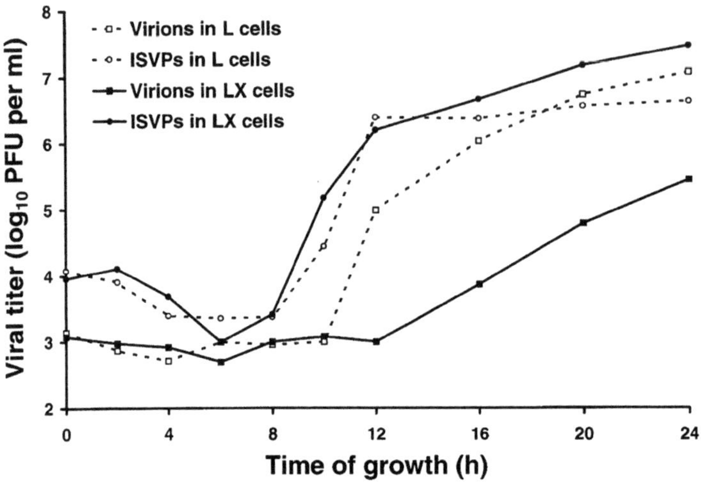

Fig. 3. Growth of wt virions and ISVPs in parental L cells and cured LX cells. Monolayers of cells (5×10^5 cells) were infected with either virions or ISVPs of wt T3D at an MOI of 2 PFU per cell. After a 1-h adsorption period the inoculum was removed, fresh medium was added, and the cells were incubated at 37°C for the times shown. Cells were frozen and thawed twice, and virus in cell lysates was titrated on L-cell monolayers by plaque assay. The results are presented as the mean viral titers for two independent experiments. (From DERMODY et al. 1993)

chloride (DERMODY et al. 1993; WETZEL et al. 1997a,b) (Fig. 4a), which suggests that mutant cells that do not fully support virion-to-ISVP processing select mutant viruses that tolerate higher pH during steps to complete viral entry. PI viruses are also capable of efficient growth in L cells treated with E64 (BAER and DERMODY 1997), an inhibitor of cysteine proteases such as those present in the endocytic compartment (BARRETT et al. 1982) (Fig. 4b). As with ammonium chloride, E64 blocks infection by virions but not ISVPs (BAER and DERMODY 1997). Thus PI viruses are altered in their requirements for both acidification and proteolysis to facilitate entry into cells.

The reovirus σ3 and μ1/μ1C proteins are major components of the viral outer capsid (SMITH et al. 1969; DRYDEN et al. 1993), and both proteins are cleaved during conversion of virions to ISVPs (CHANG and ZWEERINK 1971; SILVERSTEIN et al. 1972; STURZENBECKER et al. 1987). Since PI viruses are capable of growth in the presence of the disassembly inhibitors ammonium chloride and E64, it was reasoned that PI virus virions undergo virion-to-ISVP processing more efficiently than virions of wt virus. This hypothesis was tested in a single study in which the fate of viral structural proteins was assessed by sodium dodecyl sulfate polyacrylamide gel electrophoresis after treatment of virions of wt and PI viruses with chymotrypsin in vitro (WETZEL et al. 1997b). Proteolysis of PI virus outer-capsid proteins σ3 and μ1C occurred with faster kinetics than proteolysis of wt virus outer-capsid proteins (Fig. 5). These results provide strong evidence that increased efficiency of proteolysis of the viral outer capsid is important for growth of reovirus in persistently infected cultures.

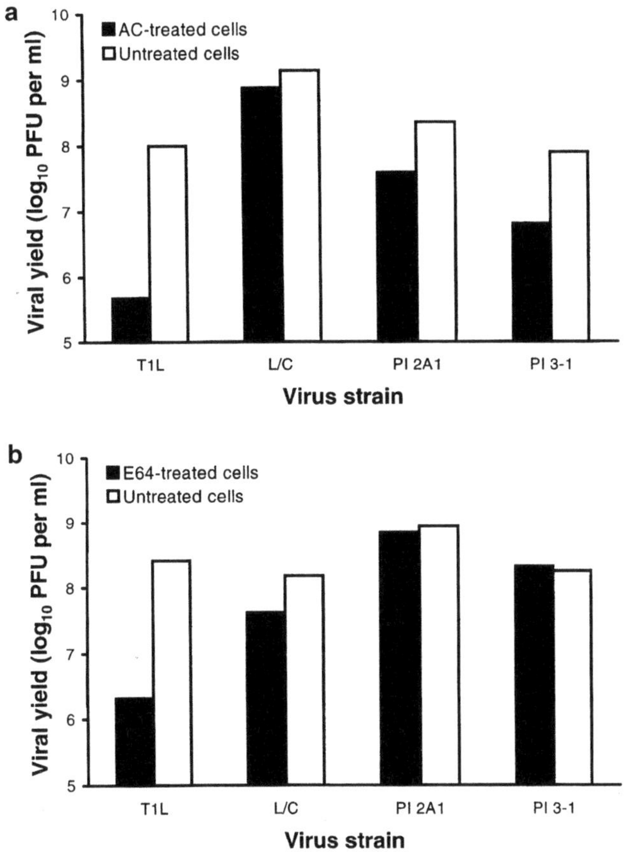

Fig. 4. Growth of wt and PI viruses in the presence and absence of (**a**) ammonium chloride (AC) and (**b**) E64. Monolayers of L cells (5×10^5cells) were infected with either wt T1L or the PI viruses shown at an MOI of 2 PFU per cell. After a 1-h adsorption period the inoculum was removed, fresh medium was added (with or without 10 mM AC or 100 μM E64), and the cells were incubated at 37°C for 10 millimolar AC 24h. Cells were frozen and thawed twice, and virus in cell lysates was titrated on L-cell monolayers by plaque assay. The results are presented as the mean viral yields 100 micromolar E64 for two independent experiments. (Adapted from Wetzel et al. 1997b; Baer and Dermody 1997)

5 A Model of Persistent Reovirus Infection of Cultured Cells

Studies described thus far make it possible to propose a general model for the establishment and maintenance of persistent reovirus infection (Fig. 6). Establishment of persistent infection appears to be associated with attenuation of viral cytopathicity, and depending on the cell type several mechanisms can facilitate this attenuation. These include viral passage at high multiplicity of infection (L cells) (Ahmed and Graham 1977; Ahmed and Fields 1982; Brown et al. 1983; Dermody et al. 1993), resistance to virus-induced inhibition of cellular protein synthesis

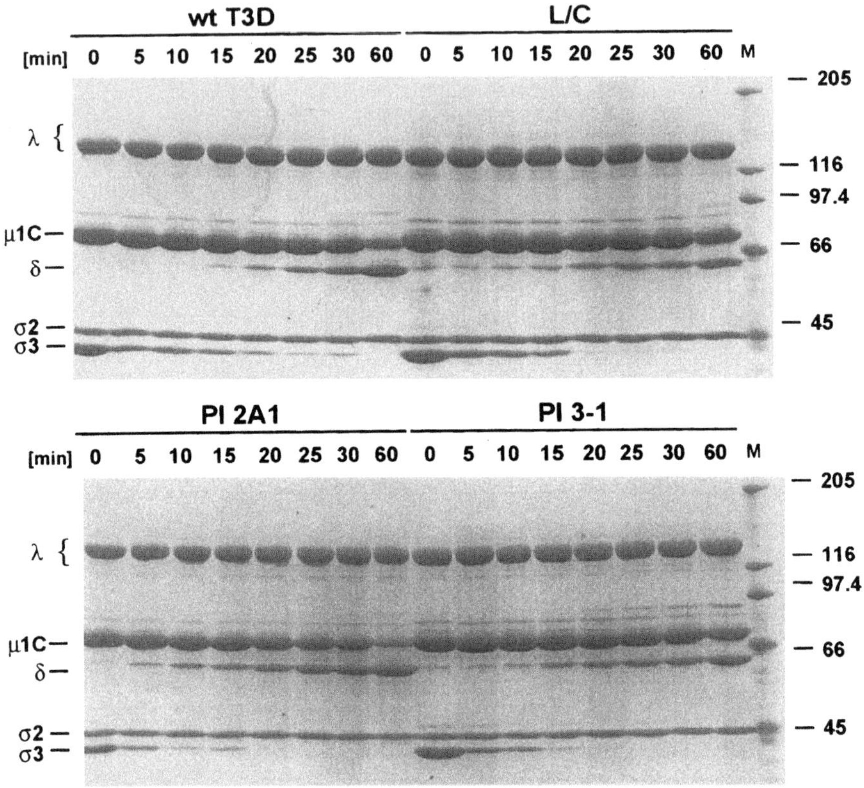

Fig. 5. Electrophoretic analysis of viral structural proteins of wt and PI virus virions during treatment with chymotrypsin to generate ISVPs. Purified virions of wt T3D and PI viruses L/C, PI 2A1, and PI 3-1 at a concentration of 8×10^{12} particles per milliliter were treated with chymotrypsin at 10°C for the times shown. Equal volumes of samples (75μl) were loaded into wells of 5%–15% polyacrylamide gradient gels. After electrophoresis gels were stained with Coomassie blue. Times (minutes) of chymotrypsin treatment are shown at the top of each gel. Viral proteins are labeled, and molecular-weight markers (in kilodaltons) appear in the lanes labeled M. (From WETZEL et al. 1997b)

(SC1 cells) (DANIS et al. 1993), and diminished capacity to support viral disassembly [L cells treated with ammonium chloride (CANNING and FIELDS 1983) and MEL cells (WETZEL et al. 1997a)]. Each of these mechanisms would result in a diminution of the effective viral inoculum, which would limit productive infection to a minority population of cells during the initial rounds of viral replication.

Cells manifesting moderate levels of resistance to viral replication would be spared and become the source for selection of an increasingly resistant cell population capable of surviving increasing viral titers. For cells with low levels of resistance to viral replication, cell crises would be expected during this period of persistent infection, leaving only those cells that support greatly reduced viral replication. During the maintenance phase of persistent infection this model predicts that mutant viruses exhibiting an augmented capacity to infect the resistant

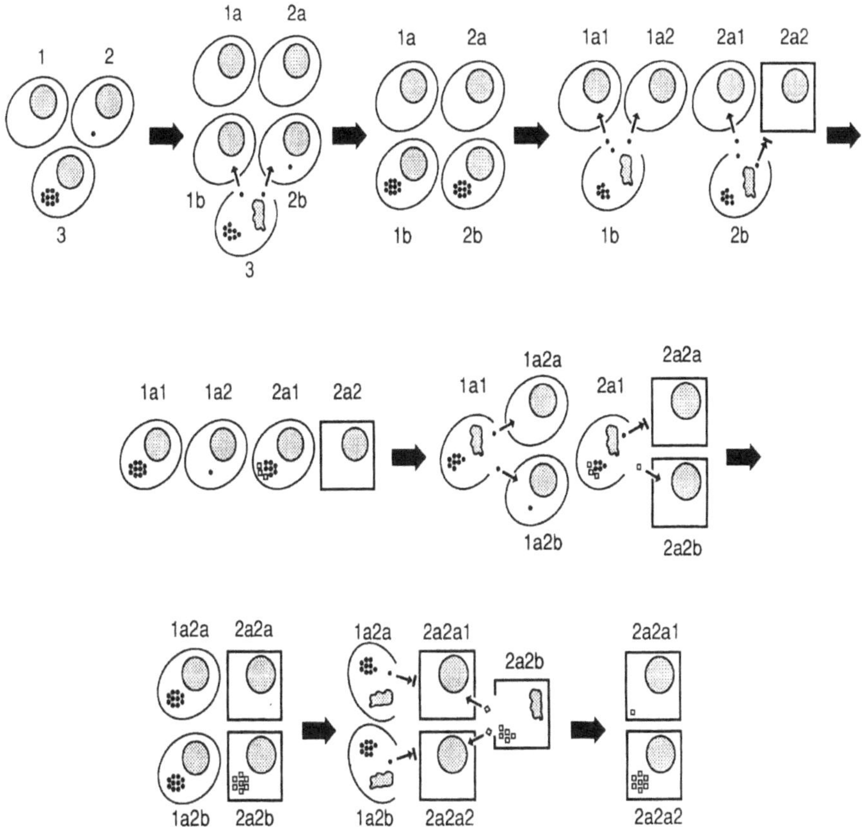

Fig. 6. Virus-cell coevolution during the maintenance of persistent reovirus infection. Cells and their progeny (large ovals and squares) and viruses and their progeny (small circles and squares) in a persistently infected culture are shown. According to this model, the culture exhibits inefficient vertical transmission of virus between cells because more severely infected cells undergo lysis (e.g., cell 3) and less severely infected cells (e.g., cell 2) undergo cell division and generate daughter cells (e.g., cells 2a and 2b) that remain susceptible to viral infection by horizontal transmission. Because of ongoing reinfection and lysis of cells within the culture, mutant cells more resistant to reinfection are selected (e.g., cell 2a2). In this model mutant viruses that are more efficient than wild-type viruses at infecting the resistant cells are selected subsequently (e.g., virus infecting cell 2a2b). In order for the persistent infection to be maintained mutant cells must not be fully permissive to infection by mutant viruses, and mutant viruses must retain the capacity to infect a subpopulation of mutant cells. (From Dermody et al. 1993)

cells would be selected. However, for these persistent infections to survive, an equilibrium between viral cytopathicity and cellular resistance must be reached in which ongoing viral replication is not sufficient to completely lyse the culture. Thus these persistent infections also can be termed chronic infections in which lysis is restricted to a subset of cells. For persistent reovirus infections of L cells and MEL cells this equilibrium rests at an early step in the viral replication cycle since these persistent infections select mutations in both cells and viruses that affect viral entry. It is possible that mutations affecting other aspects of reovirus replication are selected during persistent infections of other cell types. In this regard persistent reovirus infections of 3T3 cells are associated with decreased expression of epidermal growth factor receptors and increased expression of insulin receptors (VERDIN et al. 1986). Additionally, persistent infections of MDCK cells are associated with decreased expression of epidermal growth factor receptors and decreased capacity to form tight junctions (MONTGOMERY et al. 1991). However, the relationship of these findings to mechanisms that serve to propagate persistent infections of 3T3 cells and MDCK cells is not known.

An important conclusion from studies of persistent reovirus infections of L cells and MEL cells is that maintenance of persistent infection selects viruses with enhanced cytopathicity. This finding is in conflict with some models of persistent infection which hold that viruses with attenuated cytopathicity are required for maintenance of persistent infection. By virtue of mutations that accelerate their disassembly PI viruses selected during persistent infections of L cells and MEL cells produce larger plaques and grow to higher titers than wt viruses (J.D. WETZEL, and T.S. DERMODY, unpublished observations). Thus mutant viruses selected by growth in cells containing blocks to viral infection have an enhanced capacity to infect parental cells. It is unlikely that such virus-cell coevolution occurs in a single infected host; however, virus-host coevolution analogous to persistent reovirus infection of cultured cells is a general feature of pathogen-host interactions in populations (FENNER and KERR 1994).

6 Viral Mutations Selected During Persistent Reovirus Infection

Recent work has focused on the identification of viral mutations that confer phenotypes selected during persistent reovirus infection with the goal of determining how these mutations lead to alterations in viral entry. These studies have been greatly facilitated by reassortant genetics in which mutations responsible for entry-enhancing phenotypes can be ascribed to specific viral genes. This approach has allowed mutations important for phenotypes required to maintain persistent infection to be distinguished from irrelevant mutations arising during prolonged viral passage in cell culture. In comparison to wt viruses, PI viruses produce significantly greater yields in mutant cells cured of persistent infection (AHMED et al. 1981; KAUFFMAN et al. 1983; DERMODY et al. 1993; WILSON et al. 1996; WETZEL et al. 1997a) and in cells treated with either ammonium chloride (DERMODY et al. 1993; WETZEL et al. 1997a,b) or E64 (BAER and DERMODY 1997).

Each of these phenotypes has been mapped genetically using reassortant viruses isolated from crosses of wt strain T1L and three independent PI viruses (Table 1). Mutations in PI viruses that confer growth in cured cells segregate with either the S1 or S4 gene segments, depending on the PI virus studied (KAUFFMAN et al. 1983; WILSON et al. 1996). Similarly, mutations that confer growth in the presence of ammonium chloride segregate with either the S1 or S4 genes (WETZEL

Table 1. Viral genes that determine growth of PI viruses in mutant cells and in cells treated with either ammonium chloride or protease inhibitor E64[a]

	Viral genes that segregate with growth in		
PI virus strain	Mutant cells	AC-treated cells	E64-treated cells
L/C	S1	S1	S4
PI 2A1	S4	S4	S4
PI 3–1	S1	S4	S4

[a] PI × wt reassortant viruses isolated from three independent crosses were tested for growth in mutant cells selected during persistent infection [T1L × L/C reassortants (KAUFFMAN et al. 1983), T1L × PI 2A1 and T1L × PI 3–1 reassortants (WILSON et al. 1996)], in the presence of ammonium chloride (AC) (WETZEL et al. 1997b), and in the presence of E64 (BAER and DERMODY 1997). Genes derived from the PI virus parent that segregate with mutant viral phenotypes are shown.

et al. 1997b). These findings are provocative and suggest that acid-dependent disassembly events during conversion of virions to ISVPs involve both viral attachment protein σ1 and outer-capsid protein σ3. In contrast, mutations that confer growth in the presence of E64 map exclusively to the S4 gene segment (BAER and DERMODY 1997), which suggests that viral susceptibility to proteolytic action is determined by the σ3 protein alone.

Sequence analysis of genes that segregate with phenotypes selected during persistent infection has revealed insight into mechanisms of entry-enhancing mutations in PI viruses. The S1 gene nucleotide sequences of seven PI viruses isolated from independent persistently infected L-cell cultures have been determined (WILSON et al. 1996). The S1 sequences of these viruses contain from one to three mutations, and with a single exception each mutation results in a change in the deduced amino acid sequence of σ1 protein (Fig. 7). The capacity of the PI virus having a wt σ1 protein to efficiently infect cured cells segregates with the S4 gene and, as would be predicted, not with the S1 gene (WILSON et al. 1996). Mutations in the σ1 proteins of several PI viruses are contained in a region of σ1 important for stability of σ1 oligomers (LEONE et al. 1991b). An oligomeric form of σ1 protein of wt T3D can be detected in sodium dodecyl sulfate polyacrylamide gels by increasing the pH of the sample buffer (BASSEL-DUBY et al. 1987). Using sample buffer conditions favoring migration of σ1 oligomers, mutations in PI virus σ1 proteins were found to decrease stability of σ1 oligomers (WILSON et al. 1996) (Fig. 8). Alterations in stability of σ1 oligomers might affect conformational changes in σ1 that occur during reovirus entry. Such a conformational change might target the virus-receptor complex to an endocytic compartment where proteolysis of the outer capsid occurs or facilitate proteolysis of outer-capsid proteins by endocytic proteases. Alternatively, mutations affecting stability of σ1 oligomers might alter later steps in viral entry, such as interaction of processed ISVPs with vacuolar membranes or activation of the viral transcriptase. Enhancement of any of these entry steps would likely augment viral growth in cells manifesting blocks to viral disassembly, such as those selected during persistent reovirus infection.

The observation that oligomers of PI virus σ1 protein are less stable than wt σ1 at increased pH (WILSON et al. 1996) suggests that a conformational change in

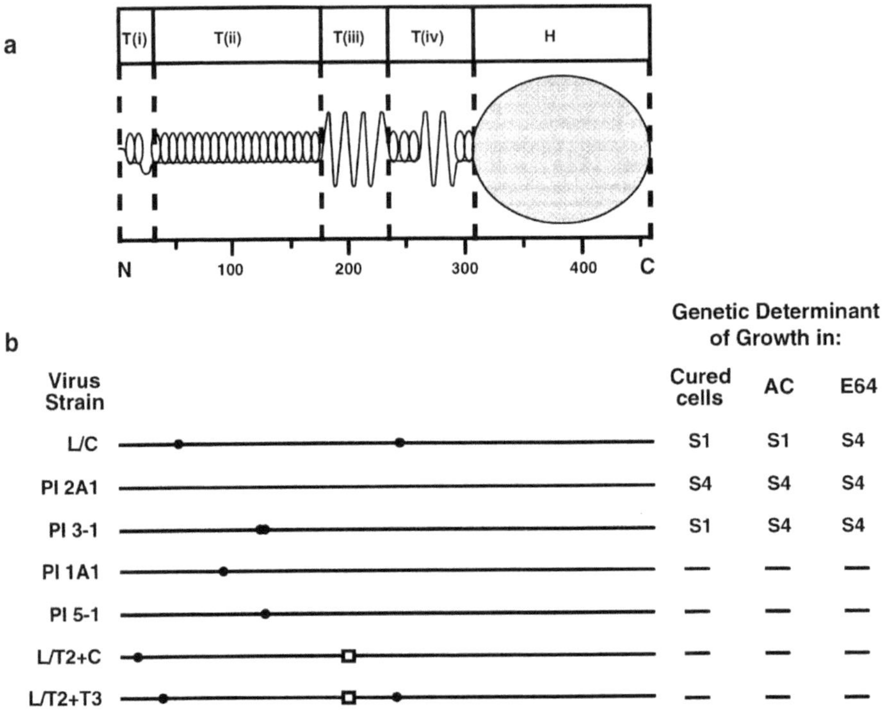

Fig. 7a,b. A structural model of the reovirus σ1 protein and location of mutations in the deduced σ1 amino acid sequences of seven PI viruses. **a** Morphological regions of σ1 (tail, *T*; head, *H*) defined by analysis of electron-microscopic images of purified σ1 (FRASER et al. 1990). The model of σ1 structure is based on analysis of deduced σ1 amino acid sequences of prototype strains of the three reovirus serotypes (NIBERT et al. 1990). The fibrous tail is proposed to be constructed from a tandem arrangement of α-helix and β-sheet; the head is predicted to assume a more complex, globular structure. Regions of α-helix and β-sheet are indicated in the σ1 tail. Amino acid positions in σ1 sequence are shown. **b** Mutations in σ1 proteins of PI viruses. *Closed circles*, sites of point mutations; *open squares*, sites of deletions. Viral genes that segregate with growth in cured cells (KAUFFMAN et al. 1983; WILSON et al. 1996), ammonium chloride (*AC*)-treated cells (WETZEL et al. 1997b), or E64-treated cells (BAER and DERMODY 1997) are indicated for PI virus strains L/C, PI 2A1, and PI 3-1; –, genetic analysis not performed. (Adapted from WILSON et al. 1996)

σ1 during viral disassembly (FURLONG et al. 1988; DRYDEN et al. 1993; NIBERT et al. 1995) is acid dependent. This contention is also supported by genetic linkage of the S1 gene and the capacity of PI virus L/C to grow in the presence of ammonium chloride (WETZEL et al. 1997b). Acid-dependent conformational changes in viral attachment proteins during disassembly have been reported for several viruses, including influenza virus (BULLOUGH et al. 1994), Semliki Forest virus (KIELIAN and HELENIUS 1985; WAHLBERG et al. 1992), and tick-borne encephalitis virus (ALLISON et al. 1995). Furthermore, pH-sensitive events involving viral attachment proteins have been shown to be altered in some types of persistent infections, including those caused by the coronavirus, mouse hepatitis virus (GALLAGHER et al. 1991).

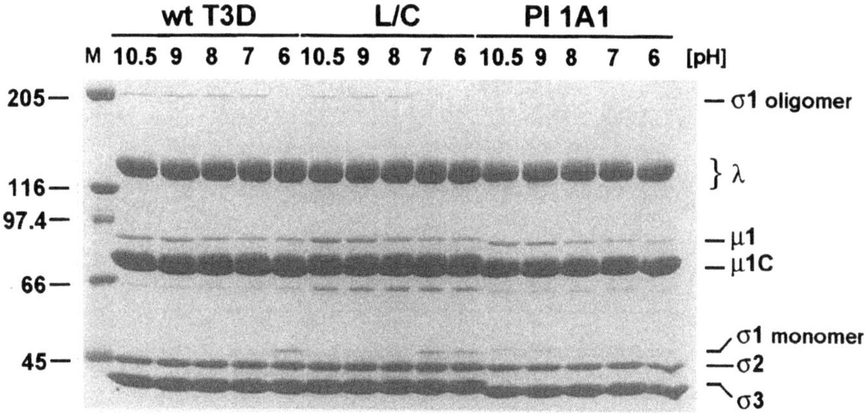

Fig. 8. Effect of pH on electrophoretic mobility of σ1 protein. Purified virions (2×10^{11} particles) of wt T3D and PI viruses L/C and PI 1A1 were incubated in sample buffer adjusted to the pH values shown prior to electrophoresis in a 5%–15% polyacrylamide gradient gel. After electrophoresis the gel was stained with Coomassie blue. Viral proteins are labeled, and molecular weight markers (in kilodaltons) appear in the lane labeled M. Oligomer and monomer bands of σ1 are indicated. (Adapted from WILSON et al. 1996)

The S4 gene nucleotide sequences of the same seven PI viruses have also been determined (WETZEL et al. 1997b). The S4 sequences of these PI viruses contain from one to four mutations, and with a single exception each mutation results in a substitution in the deduced amino acid sequence of σ3 protein (Fig. 9). Three regions of σ3 appear to be targets for mutations in the PI viruses studied: amino acids 86–145, 218–232, and 354. Six of the seven PI viruses studied contain a tyrosine to histidine substitution at residue 354, and in the case of PI virus 3-1 σ3 protein, this substitution is the only mutation observed. Since the S4 gene segregates exclusively with the growth of T1L × PI 3-1 reassortants in cells treated with either ammonium chloride (WETZEL et al. 1997b) or E64 (BAER and DERMODY 1997), it appears that a tyrosine to histidine mutation at amino acid 354 determines susceptibility of the σ3 protein to acid-dependent proteolysis. A region of σ3 adjacent to amino acid 220 is sensitive to a variety of proteases (SCHIFF et al. 1988; MILLER and SAMUEL 1992), and this region of the protein is postulated to be cleaved by endocytic proteases during viral entry (SHEPARD et al. 1995). It is possible that the tyrosine to histidine mutation at amino acid 354 alters the conformation of the σ3 cleavage site and enhances susceptibility of σ3 to proteolysis by E64-sensitive proteases. Alternatively, this mutation might alter interactions between σ3 and another outer-capsid protein such that the σ3 cleavage site is indirectly rendered more accessible to proteolysis. In support of this idea, it has been shown that interactions between σ3 and μ1 result in a conformational change in σ3 that increases its susceptibility to cleavage (SHEPARD et al. 1995). Another possibility is that the tyrosine to histidine mutation at amino acid 354 allows σ3 to be cleaved by acid-independent proteases that are not inhibited by E64.

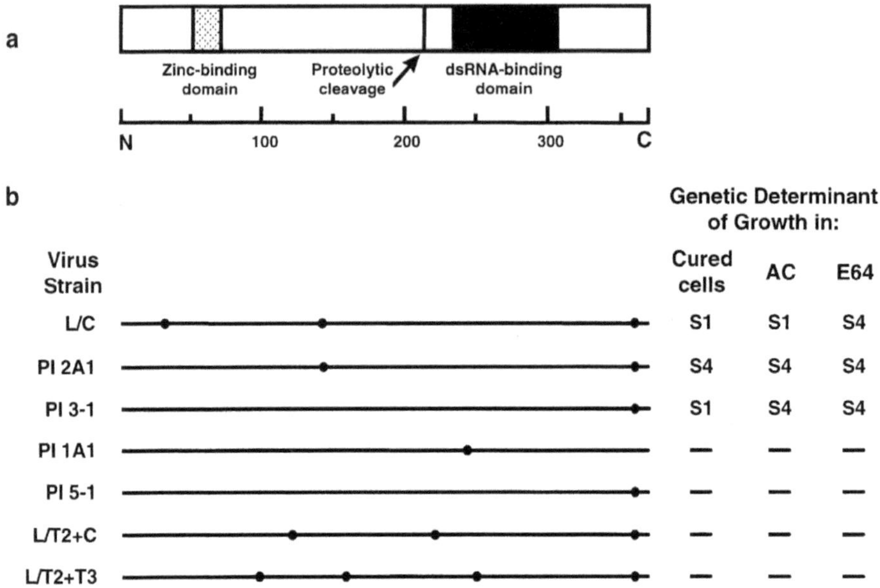

Fig. 9a,b. Location of mutations in deduced σ3 amino acid sequences of seven PI viruses. **a** Functional domains of σ3 protein (SCHIFF et al. 1988; MILLER and SAMUEL 1992). *Arrow*, a site cleaved by staphylococcal V8 protease (SCHIFF et al. 1988). Amino acid positions in σ3 sequence are shown. **b** Mutations in σ3 proteins of PI viruses. Viral genes that segregate with growth in cured cells (KAUFFMAN et al. 1983; WILSON et al. 1996), ammonium chloride (*AC*)-treated cells (WETZEL et al. 1997b), or E64-treated cells (BAER and DERMODY 1997) are indicated for PI virus strains L/C, PI 2A1, and PI 3-1; –, genetic analysis not performed. (Adapted from WETZEL et al. 1997b)

7 A Model of Reovirus Entry Derived from Studies of PI Reoviruses

The identification of viral genes that segregate with PI virus growth in cells treated with either ammonium chloride or E64 has led to the suggestion that acidification and proteolysis mediate different events in reovirus disassembly. Growth of T1L × L/C reassortants in the presence of ammonium chloride segregates with the S1 gene (WETZEL et al. 1997b), whereas growth of T1L × L/C reassortants in the presence of E64 segregates with the S4 gene (BAER and DERMODY 1997). Thus in the case of L/C, mutations in σ1 affect acid-dependent disassembly steps, and mutations in σ3 affect protease-dependent disassembly steps. The temporal relationship between these processes was tested by adding either ammonium chloride or E64d, a membrane permeable form of E64, at various times after viral adsorption (BAER and DERMODY 1997). Reovirus growth was found to be susceptible to complete blockade by both ammonium chloride and E64 only up to 30min after viral adsorption; thereafter, susceptibility to both inhibitors decreased logarithmically for an additional 30 min. At times of addition greater than 60min after adsorption, neither ammonium chloride nor E64 had a significant effect on reovirus growth. Therefore these results suggest that acid-dependent and protease-dependent events are independent but temporally associated steps in reovirus disassembly and likely occur within the same cellular compartment.

Studies of reovirus mutants selected during persistent infection have identified viral structural proteins that mediate requirements for acidification and proteolysis during viral entry. Additionally, these studies have established the molecular basis for viral resistance to inhibitors of virion-to-ISVP disassembly. Based on this work, current models of reovirus entry can be revised to incorporate independent roles of acidification and proteolysis during disassembly of reovirus virions (Fig. 10). Reovirus disassembly is likely initiated by acid-dependent processes involving the σ1 and σ3 proteins, which are followed rapidly by proteolysis of σ3. These events are in turn followed by proteolysis of μ1/μ1C to yield the fully processed ISVP, which is capable of membrane penetration. Studies of PI reoviruses have identified the σ1 and σ3 proteins as critical targets for mutations that enhance reovirus entry. It is noteworthy that mutations in either protein facilitate growth in cured cells and resistance to ammonium chloride. These observations suggest that the σ1 and σ3 proteins interact to facilitate disassembly of reovirus virions, perhaps by mediating acid-dependent conformational changes required for subsequent proteolysis of the viral outer capsid.

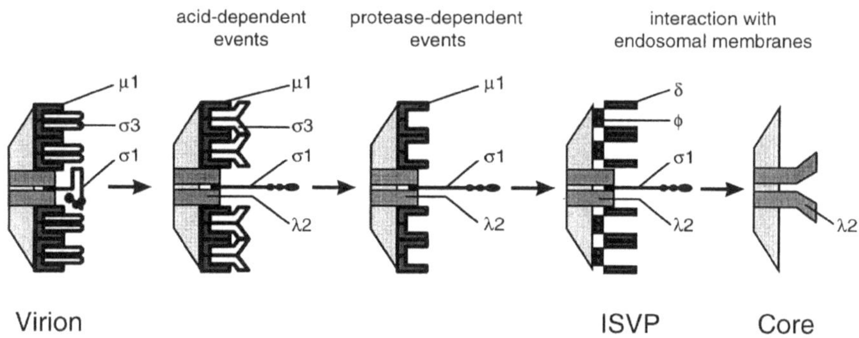

Fig. 10. A model of the disassembly of reovirus virions derived from studies of PI reoviruses. After reovirus is taken into cells by receptor-mediated endocytosis, disassembly of reovirus virions occurs in cellular endosomes by ordered changes in the viral outer capsid. Shown here are disassembly events affecting a single viral vertex. First, acid-dependent conformational changes occur in viral attachment protein σ1 and outer-capsid protein σ3. Second, the σ3 protein is degraded by endocytic proteases and lost from virions. Third, outer-capsid protein μ1/μ1C is cleaved to form μ1δ/δ and ϕ, yielding the fully processed ISVP, which is capable of interacting with endosomal membranes. The transcriptionally active viral core does not contain outer-capsid proteins and demonstrates extensive conformational changes in core-spike protein λ2. (DRYDEN et al. 1993)

8 Pathogenesis of PI Reoviruses

The finding that PI reoviruses are altered in viral entry led to an examination of whether these viruses are altered in virulence (MORRISON et al. 1993). Newborn NIH-Swiss mice were inoculated intracranially with 16 PI viruses, and LD_{50} values were determined for each strain. Twelve of these PI viruses had LD_{50} values identical to that of wt T3D, and four were attenuated. Of mice surviving inoculation with PI viruses at doses corresponding to the LD_{50}, 38% had detectable virus in brain tissue 25 days following infection, and of these one-half had titers greater than 1×10^5PFU. In contrast, only 16% of mice inoculated with T3D had detectable viral titer in brain tissue on day 25, and no titer was greater than 10^5PFU. Tropism of PI virus within the brain resembled that of wt virus, and the distribution of PI virus antigen in brain tissue did not change over time. By 50 days after intracranial inoculation with either wt or PI virus none of the surviving mice had detectable viral titer in brain tissue. These observations suggest that entry-enhancing mutations in PI viruses do not significantly alter viral virulence, but they do lead to prolonged viral replication in vivo.

Although reoviruses do not establish persistent infections of immunocompetent mice, they can cause persistent infections of severe combined immunodeficiency (SCID) mice (HALLER et al. 1995). Adult SCID mice inoculated intraperitoneally with wt T3D were found to survive for periods of 100 days or longer. Surviving animals harbored virus in a variety of tissues, including brain, liver, and spleen, and some of these viruses were adapted to enhanced growth in organs from which they were isolated. None of the organ-specific variants could grow in the presence of ammonium chloride, which suggests that mutations affecting acid-dependent entry steps are not selected during persistent infections in vivo. Nonetheless, the general principle of selection of viral variants dependent on the host cell is applicable to this model of persistent infection, and it is likely that analysis of variants adapted to growth in particular host tissues will reveal important new information about cell-specific factors required for efficient viral replication.

9 Future Prospects

Studies of persistent reovirus infections of cultured cells show that the cytolytic potential of mutant viruses is not diminished and that mutations in both viruses and cells affect early steps in reovirus replication involving acid-dependent viral disassembly in cellular endosomes (DERMODY et al. 1993; WETZEL et al. 1997a). These mutant viruses and cells represent a natural perturbation of the entry process and offer a unique opportunity to define mechanisms by which reovirus enters cells. Important directions for future research include a precise determination of the cellular mutations that lead to selection of mutant viruses altered in viral entry and characterization of the molecular and structural basis of the disassembly process. Since basic mechanisms of cell entry are not well understood for most nonenveloped viruses, studies of the interplay between viral and cellular factors required to effect reovirus entry should contribute significantly to this field and illuminate new targets for therapeutic intervention in diseases caused by viruses that enter cells by receptor-mediated endocytosis. Moreover, ongoing studies of adaptive mechanisms used by viruses to overcome host-cell barriers to infection will reveal critical balance points in viral evolution and highlight strategies used by viruses to infect naive host populations.

Acknowledgements. I thank Erik Barton, Geoff Baer, Jim Chappell, Denise Wetzel, and Greg Wilson for careful review of the manuscript. This work was supported by Public Health Service awards AI32539 and AI38296 from the National Institute of Allergy and Infectious Diseases and the Elizabeth B. Lamb Center for Pediatric Research.

References

Ahmed R, Fields BN (1982) Role of the S4 gene in the establishment of persistent reovirus infection in L cells. Cell 28:605–612

Ahmed R, Graham AF (1977) Persistent infections in L cells with temperature-sensitive mutants of reovirus. J Virol 23:250–262

Ahmed R, Chakraborty PR, Graham AF, Ramig RF, Fields BN (1980) Genetic variation during persistent reovirus infection: presence of extragenically suppressed temperature-sensitive lesions in wild-type virus isolated from persistently infected L cells. J Virol 34:383–389

Ahmed R, Canning WM, Kauffman RS, Sharpe AH, Hallum JV, Fields BN (1981) Role of the host cell in persistent viral infection: coevolution of L cells and reovirus during persistent infection. Cell 25:325–332

Ahmed R, Kauffman RS, Fields BN (1983) Genetic variation during persistent reovirus infection: isolation of cold-sensitive and temperature-sensitive mutants from persistently infected L cells. Virology 131:71–78

Ahmed R, Morrison LA, Knipe DM (1996) Persistence of viruses. In: Fields BN, Knipe DM, Howley PM (eds) Fields virology. Lippincott-Raven, Philadelphia, pp 219–249

Allison SL, Schalich J, Stiansy K, Mandl CW, Kunz C, Heinz FX (1995) Oligomeric rearrangement of tick-borne encephalitis virus envelope proteins induced by an acidic pH. J Virol 69:695–700

Baer GS, Dermody TS (1997) Mutations in reovirus outer-capsid protein σ3 selected during persistent infections of L cells confer resistance to protease inhibitor E64. J Virol 71:4921–4928

Barrett AJ, Kembhavi AA, Brown MA, Kirschke H, Knight CG, Tamai M, Hanada K (1982) L-*trans*-Epoxysuccinyl-leucylamido(4-guanidino)butane (E-64) and its analogues as inhibitors of cysteine proteinases including cathepsins B, H and L. Biochem J 201:189–198

Bass DM, Bodkin D, Dambrauskas R, Trier JS, Fields BN, Wolf JL (1990) Intraluminal proteolytic activation plays an important role in replication of type 1 reovirus in the intestines of neonatal mice. J Virol 64:1830–1833

Bassel-Duby R, Nibert M, Homcy C, Fields B, Sawutz D (1987) Evidence that the sigma 1 protein of reovirus serotype 3 is a multimer. J Virol 61:1834–1841

Bell TM, Massie A, Ross MGR (1966) Further isolation of reovirus type 3 from cases of Burkitt's lymphoma. Br Med J 2:1514–1517

Bodkin DK, Nibert ML, Fields BN (1989) Proteolytic digestion of reovirus in the intestinal lumens of neonatal mice. J Virol 63:4676–4681

Borsa J, Morash BD, Sargent MD, Copps TP, Lievaart PA, Szekely JG (1979) Two modes of entry of reovirus particles into L cells. J Gen Virol 45:161–170

Borsa J, Sargent MD, Lievaart PA, Copps TP (1981) Reovirus: evidence for a second step in the intracellular uncoating and transcriptase activation process. Virology 111:191–200
Borzakian S, Couderc T, Barbier Y, Attal G, Pelletier I, Colbere-Garapin F (1992) Persistent poliovirus infection: establishment and maintenance involve distinct mechanisms. Virology 186:398–408
Brown EG, Nibert ML, Fields BN (1983) The L2 gene of reovirus serotype 3 controls the capacity to interfere, accumulate deletions and establish persistent infection. In: Compans RW, Bishop DHL (eds) Double-stranded RNA viruses. Elsevier Biomedical, New York, pp 275–287
Bullough PA, Hughson FM, Skehel JJ, Wiley DC (1994) Structure of influenza haemagglutinin at the pH of membrane fusion. Nature 371:37–43
Canning WM, Fields BN (1983) Ammonium chloride prevents lytic growth of reovirus and helps to establish persistent infection in mouse L cells. Science 219:987–988
Chang CT, Zweerink HJ (1971) Fate of parental reovirus in infected cell. Virology 46:544–555
Chen W, Baric RS (1996) Molecular anatomy of mouse hepatitis virus persistence: coevolution of increased host cell resistance and virus virulence. J Virol 70:3947–3960
Cleveland DR, Zarbl H, Millward S (1986) Reovirus guanyltransferase is L2 gene product lambda 2. J Virol 60:307–311
Danis C, Mabrouk T, Garzon S, Lemay G (1993) Establishment of persistent reovirus infection in SC1 cells: absence of protein synthesis inhibition and increased level of double-stranded RNA-activated protein kinase. Virus Res 2727:253–265
Dermody TS, Nibert ML, Wetzel JD, Tong X, Fields BN (1993) Cells and viruses with mutations affecting viral entry are selected during persistent infections of L cells with mammalian reoviruses. J Virol 67:2055–2063
Dermody TS, Chappell JD, Hofler JG, Kramp W, Tyler KL (1995) Eradication of persistent reovirus infection from a B-cell hybridoma. Virology 212:272–276
Dryden KA, Wang G, Yeager M, Nibert ML, Coombs KM, Furlong DB, Fields BN, Baker TS (1993) Early steps in reovirus infection are associated with dramatic changes in supramolecular structure and protein conformation: analysis of virions and subviral particles by cryoelectron microscopy and image reconstruction. J Cell Biol 122:1023–1041
Duncan MR, Stanish SM, Cox DC (1978) Differential sensitivity of normal and transformed human cells to reovirus infection. J Virol 28:444–449
Fenner F, Kerr PJ (1994) Evolution of the poxviruses, including the coevolution of virus and host in myxomatosis In: Morse SS (ed) The evolutionary biology of viruses. Raven, New York, pp 273–292
Fraser RD, Furlong DB, Trus BL, Nibert ML, Fields BN, Steven AC (1990) Molecular structure of the cell-attachment protein of reovirus: correlation of computer-processed electron micrographs with sequence-based predictions. J Virol 64:2990–3000
Furlong DB, Nibert ML, Fields BN (1988) Sigma 1 protein of mammalian reoviruses extends from the surfaces of viral particles. J Virol 62:246–256
Gallagher TM, Escarmis C, Buchmeier MJ (1991) Alteration of the pH dependence of coronavirus-induced cell fusion: effect of mutations in the spike glycoprotein. J Virol 65:1916–1928
Haller BL, Barkon ML, Li X-Y, Hu WM, Wetzel JD, Dermody TS, Virgin HW (1995) Brain- and intestine-specific variants of reovirus serotype 3 strain Dearing are selected during chronic infection of severe combined immunodeficient mice. J Virol 69:3933–3937
Kaplan G, Levy A, Racaniello VR (1989) Isolation and characterization of HeLa cell lines blocked at different steps in the poliovirus life cycle. J Virol 63:43–51
Kauffman RS, Ahmed R, Fields BN (1983) Selection of a mutant S1 gene during reovirus persistent infection of L cells: role in maintenance of the persistent state. Virology 131:79–87
Kielian MC, Helenius A (1985) pH-induced alterations in the fusogenic spike protein of Semliki Forest virus. J Cell Biol 101:2284–2291
Lee PW, Hayes EC, Joklik WK (1981) Protein sigma 1 is the reovirus cell attachment protein. Virology 108:156–163
Leone G, Duncan R, Lee PWK (1991a) Trimerization of the reovirus cell attachment protein ($\sigma 1$) induces conformational changes in $\sigma 1$ necessary for its cell-binding function. Virology 184:758–761
Leone G, Duncan R, Mah DC, Price A, Cashdollar LW, Lee PW (1991b) The N-terminal heptad repeat region of reovirus cell attachment protein $\sigma 1$ is responsible for $\sigma 1$ oligomer stability and possesses intrinsic oligomerization function. Virology 182:336–345
Leone G, Maybaum L, Lee PWK (1992) The reovirus cell attachment protein possesses two independently active trimerization domains: basis of dominant negative effects. Cell 71:479–488
Levy JA, Henle G, Henle W, Zajac BA (1968) Effect of reovirus type 3 on cultured Burkitt's tumour cells. Nature 220:607–608

Mahy BWJ (1985) Strategies of viral persistence. Br Med Bull 41:50–55
Martinez CG, Guinea R, Benavente J, Carrasco L (1996) The entry of reovirus into L cells is dependent on vacuolar proton-ATPase activity. J Virol 70:576–579
Matsuzaki N, Hinshaw VS, Fields BN, Greene MI (1986) Cell receptors for the mammalian reovirus: reovirus-specific T-cell hybridomas can become persistently infected and undergo autoimmune stimulation. J Virol 60:259–266
Merion ML, Schlesinger P, Brooks RM, Moehring JM, Moehring TJ, Sly WS (1983) Defective acidification of endosomes in Chinese hamster ovary cell mutants cross-resistant to toxins and viruses. Proc Natl Acad Sci USA 80:5315–5319
Miller JE, Samuel CE (1992) Proteolytic cleavage of the reovirus sigma 3 protein results in enhanced double-stranded RNA-binding activity: identification of a repeated basic amino acid motif within the C-terminal binding region. J Virol 66:5347–5356
Montgomery LB, Kao CYY, Verdin EM, Cahill C, Maratos-Flier E (1991) Infection of a polarized epithelial cell line with wild-type reovirus leads to virus persistence and altered cellular function. J Gen Virol 72:2939–2946
Morrison LA, Fields BN, Dermody TS (1993) Prolonged replication in the mouse central nervous system of type 3 reoviruses isolated from persistently infected L-cell cultures. J Virol 67:3019–3026
Mrukowicz J, Wetzel JD, Goral MI, Fogo AB, Wright PF, Dermody TS (1997) Viruses and cells with mutations affecting viral entry are selected during persistent rotavirus infections of MA104 cells. J Virol 72 (in press)
Nibert ML, Dermody TS, Fields BN (1990) Structure of the reovirus cell-attachment protein: a model for the domain organization of σ1. J Virol 64:2976–2989
Nibert ML, Chappell JD, Dermody TS (1995) Infectious subvirion particles of reovirus type 3 Dearing exhibit a loss in infectivity and contain a cleaved σ1 protein. J Virol 69:5057–5067
Nibert ML, Schiff LA, Fields BN (1996) Reoviruses and their replication. In: Fields BN, Knipe DM, Howley PM (eds) Fields virology, 3rd edn. Lippincott-Raven, Philadelphia, pp 1557–1596
Ralph SJ, Harvey JD, Bellamy AR (1980) Subunit structure of the reovirus spike. J Virol 36:894–896
Rubin DH, Weiner DB, Dworkin C, Greene MI, Maul GG, Williams WV (1992) Receptor utilization by reovirus type 3: distinct binding sites on thymoma and fibroblast cell lines result in differential compartmentalization of virions. Microb Pathog 12:351–365
Schiff LA, Nibert ML, Co MS, Brown EG, Fields BN (1988) Distinct binding sites for zinc and double-stranded RNA in the reovirus outer capsid protein sigma 3. Mol Cell Biol 8:273–283
Sharpe AH, Fields BN (1983) Pathogenesis of reovirus infection. In: Joklik WK (eds) The reoviridae. Plenum, New York, pp 229–285
Shepard DA, Ehnstrom JG, Schiff LA (1995) Association of reovirus outer capsid proteins σ3 and μ1 causes a conformational change that renders σ3 protease sensitive. J Virol 69:8180–8184
Silverstein SC, Astell C, Levin DH, Schonberg M, Acs G (1972) The mechanism of reovirus uncoating and gene activation in vivo. Virology 47:797–806
Smith RE, Zweerink HJ, Joklik WK (1969) Polypeptide components of virions, top component and cores of reovirus type 3. Virology 39:791–810
Strong JE, Leone G, Duncan R, Sharma RK, Lee PW (1991) Biochemical and biophysical characterization of the reovirus cell attachment protein sigma 1: evidence that it is a homotrimer. Virology 184:23–32
Sturzenbecker LJ, Nibert M, Furlong D, Fields BN (1987) Intracellular digestion of reovirus particles requires a low pH and is an essential step in the viral infectious cycle. J Virol 61:2351–2361
Taber R, Alexander V, Whitford W (1976) Persistent reovirus infection of CHO cells resulting in virus resistance. J Virol 17:513–525
Verdin EM, Maratos-Flier E, Carpentier JL, Kahn CR (1986) Persistent infection with a nontransforming RNA virus leads to impaired growth factor receptors and response. J Cell Physiol 128:457–465
Wahlberg JM, Bron R, Wischut J, Garoff H (1992) Membrane fusion of Semliki Forest virus involves homotrimers of the fusion protein. J Virol 66:7309–7318
Weiner HL, Ault KA, Fields BN (1980) Interaction of reovirus with cell surface receptors. I. Murine and human lymphocytes have a receptor for the hemagglutinin of reovirus type 3. J Immunol 124:2143–2148
Wetzel JD, Chappell JD, Fogo AB, Dermody TS (1997a) Efficiency of viral entry determines the capacity of murine erythroleukemia cells to support persistent infections by mammalian reoviruses. J Virol 71:299–306

Wetzel JD, Wilson GJ, Baer GS, Dunnigan LR, Wright JP, Tang DSH, Dermody TS (1997b) Reovirus variants selected during persistent infections of L cells contain mutations in the viral S1 and S4 genes and are altered in viral disassembly. J Virol 71:1362–1369
Whitton JL, Oldstone MBA (1996) Immune responses to viruses. In: Fields BN, Knipe DM, Howley PM (eds) Fields virology. Lippincott-Raven, Philadelphia, pp 345–374
Wilson GJ, Wetzel JD, Puryear W, Bassel-Duby R, Dermody TS (1996) Persistent reovirus infections of L cells select mutations in viral attachment protein σ1 that alter oligomer stability. J Virol 70:6598–6606

Apoptosis and the Cytopathic Effects of Reovirus

S.M. OBERHAUS,[1*] T.S. DERMODY,[2] and K.L. TYLER[3]

[1] Department of Neurology, University of Colorado Health Sciences Center, Denver, CO 80262, USA
* Current Address: Department of Microbiology and Immunology, East Carolina University School of Medicine, Greenville, NC 27858, USA (E-mail: soberhaus@brody.med.ecu.edu)
[2] Departments of Pediatrics, and Microbiology and Immunology and Elizabeth B. Lamb Center for Pediatric Research, Vanderbilt University School of Medicine, Nashville, TN 37232, USA
(E-mail: terry.dermody@mcmail.vanderbilt.edu)
[3] Departments of Neurology, and Microbiology, Immunology, and Medicine, University of Colorado Health Sciences Center, 4200 East 9th Ave., Denver, CO 80262, USA
Neurology Service, Denver Veterans Affairs Medical Center, 1055 Clermont St., Denver, CO 80220, USA
(E-mail: ken.tyler@uchsc.edu)

1 Introduction

Reoviruses are cytolytic viruses which cause the death of infected cells in vitro and in vivo. The relationship between events involved in the production of virus and death of the infected cell is not clearly understood. Studies of virus-induced biochemical and morphological changes in cells, i.e., cytopathic effects, have begun to define both viral and cellular factors involved in virus-induced cell death. The description and definition of apoptosis as a mechanism of cell death distinct from necrosis has provided a framework for elucidating the roles of these viral and cellular factors in the cell death process. The demonstration that reovirus induces apoptosis in vitro and in vivo identifies apoptosis as an important component of reovirus-induced cytopathic effects and tissue injury. In this chapter we review the current data describing reovirus-induced apoptosis and propose that many of the previously described cytopathic effects of reovirus infection represent events which initiate or comprise steps in the apoptotic process.

2 Cytopathic Effects of Reovirus Infection

Reoviruses are nonenveloped viruses which bind to susceptible cells via the outer capsid protein, $\sigma 1$, and enter cells by receptor-mediated endocytosis (see NIBERT et al. 1996 for review). Virus particles are proteolytically processed within late endosomes or lysosomes and released as transcriptionally active core particles into the cytoplasm. The production of viral messenger and genomic RNA occurs within these core particles. Mature virions are released inefficiently from infected cells following cell lysis, so that some infectious virus remains associated with cellular material.

As with many viruses, reovirus infection and replication have dramatic effects on infected cells. Inhibition of host cell DNA synthesis is one of the earliest detectable effects observed 8–12h after infection with type 3 (T3) reovirus strains, concomitant with the logarithmic phase of viral growth (DUNCAN et al. 1978; ENSMINGER and TAMM 1969a; GOMATOS and TAMM 1963; SHARPE and FIELDS 1981). Previous studies indicate that DNA synthesis is inhibited during initiation of new strands and that transcription and protein synthesis are not significantly affected (ENSMINGER and TAMM 1969b; HAND et al. 1971). As a result, T3-infected cells are blocked from progressing to S phase and remain in G_1. This interruption of the cell cycle adversely affects cell metabolism and balanced growth and is associated with changes in chromatin structure characteristic of apoptotic cells (CHALY et al. 1980). UV-irradiated, replication-incompetent virus particles, but not core particles or virions without genomes, also inhibit DNA synthesis (HAND and TAMM 1973; LAI et al. 1973; SHARPE and FIELDS 1981; SHAW and COX 1973). Other prototype reovirus strains, T1 Lang (T1L) and T2 Jones (T2J), do not inhibit DNA synthesis. Using reassortant viruses, the strain-specific difference in DNA synthesis inhibition has been mapped to the S1 gene segment which encodes the σ1 and σ1s proteins (SHARPE and FIELDS 1981). Since σ1s is a nonstructural protein which is absent and cannot be synthesized by UV-irradiated preparations of purified virus, the capacity to inhibit DNA synthesis is primarily determined by σ1 alone. This idea is supported by the findings that DNA synthesis can also be inhibited by purified σ1 protein and by a mouse monoclonal anti-reovirus T3 receptor antibody (GAULTON and GREENE 1989; SARAGOVI et al. 1995).

T2J inhibits both RNA and protein synthesis, capacities primarily determined by the S4 gene which encodes the major outer capsid protein, σ3 (SHARPE and FIELDS 1982). In this case replication-competent virus is required for inhibition. Interestingly, σ3 protein is removed at one of the earliest steps of virus disassembly and presumably released into the lysosomal vacuole where proteolytic processing occurs. σ3, which has been shown to bind RNA, may inhibit RNA synthesis by binding to cellular RNA molecules (SHARPE and FIELDS 1982). T1L does not inhibit DNA or protein synthesis, or does so to a much lesser extent than T3 strains or T2J, respectively, and its effect on RNA synthesis is unknown. Infection with any one of the three prototype reovirus strains results in death of a variety of cultured cells, suggesting that cytocidal effects are not linked to synthesis inhibition of the same macromolecules.

The production of reovirus particles involves cytoskeletal elements of host cells which may disrupt normal cell structure and produce the morphological changes observed in dying, infected cells. Vimentin (intermediate) filaments are incorporated into reovirus "factories" and undergo progressive disruption and reorganization, which may lead to loss of normal cell architecture and ultimately death (SHARPE et al. 1982).

Reovirus has been shown to induce production of the cytokines interferon and tumor necrosis factor-alpha (FARONE et al. 1993; GAUNTT 1973; HENDERSON and JOKLIK 1978; LAI and JOKLIK 1973; LONG and BURKE 1971). UV-inactivated virus retains this capacity, indicating that viral replication is not required for cytokine induction (HENDERSON and JOKLIK 1978; LAI and JOKLIK 1973). Reovirus strains T1L and T3D differ in both the amount of interferon induced and sensitivity to the antiviral effects of interferon (JACOBS and FERGUSON 1991). The antiviral effects of interferon may include inhibition of viral replication and/or death of the infected cell. The induction of cytokines by reovirus can therefore be considered a cytopathic effect since infected cells are often eliminated via cytokine-mediated mechanisms, for example, removal by macrophages or natural killer cells, or by induction of apoptosis.

The lytic nature of reovirus infection and the specific cytopathic effects of this virus on its host cell suggested to us that reovirus-induced cell death occurs by apoptosis.

3 Apoptosis

Apoptosis describes cell death by a specific process dictated by cellular factors comprising a cell death "apparatus" which appears to be remarkably conserved evolutionarily. This process can be induced by a variety of stimuli, but apoptotic cells are characterized by morphological and biochemical changes which distinguish them from living cells and cells dying via nonapoptotic mechanisms (for review see MAJNO and JORIS 1995; SCHWARTZMAN and CIDLOWSKI 1993). These limited and conserved cellular changes result from the activation and expression of a cadre of cellular factors coordinated to bring about the immunologically "quiet" death of single cells. Apoptosis occurs during normal developmental and physiological processes to eliminate cells which are no longer useful or have sustained genetic damage (BURSCH et al. 1992; KERR et al. 1972). Under these conditions apoptosis serves to benefit the organism by maintaining homeostasis and balanced growth. Apoptosis also occurs under conditions which are detrimental to the organism resulting in dysregulated cell growth and death. An insufficient rate or the absence of apoptosis has been implicated in carcinogenesis and autoimmune disease. The induction of inappropriate or excessive apoptosis can produce significant tissue damage during ischemia, viral infection, and neurodegenerative diseases such as Alzheimer's disease and Parkinson's disease (GORMAN et al. 1996; THOMPSON 1995).

Morphological characteristics used to identify apoptotic cells include condensation and margination of chromatin against the nuclear membrane, loss of cytoplasmic volume with compaction of organelles, and budding of the plasma membrane (for review see MAJNO and JORIS 1995). Organelles generally remain intact, but the nucleus often breaks up into membrane-bound bodies containing chromatin. The cell itself may break apart into multiple membrane-bound apoptotic bodies containing nuclear fragments, or collapse into a single, dense, round apoptotic body. Morphological assays for apoptotic cells generally involve DNA staining and microscopic analysis for chromatin condensation, cell shrinkage, or fission into apoptotic bodies. Flow cytometry can also be used to sort apoptotic from nonapoptotic cells based on differences in size.

Biochemical changes which are consistently associated with apoptosis include specific patterns of DNA, ribosomal RNA, and protein degradation, and an increase in intracellular Ca^{2+} levels. DNA is cleaved by an endogenous endonuclease at internucleosomal sites generating fragments varying in length by multiples of 180–200bp (COHEN and DUKE 1984; DUKE et al. 1983). Single-strand nicks and fragmentation into 50–200kbp lengths may also occur before or without the generation of oligonucleosomes (WYLLIE et al. 1984). Cleavage of 28S, but not 18S ribosomal RNA has been observed to occur in several cell lines in which internucleosomal cleavage of DNA was also detected (HOUGE et al. 1995). Biochemical assays for apoptotic cells generally involve detection of 50–200kbp or oligonucleosomal-length DNA fragments using pulsed field or standard agarose gel electrophoresis and analysis by ethidium bromide staining. Apoptosis induced by a variety of intracellular and extracellular stimuli is generally characterized by one or more of the morphological and biochemical changes described but may vary with cell type, induction stimulus, and many other factors. It is therefore important to analyze cells using several different assays for both morphological and biochemical changes associated with apoptosis. Cells which are dying but do not exhibit the typical changes attributed to apoptosis are often referred to as necrotic. This terminology is commonly used to refer to nonapoptotic cells, but technically it encompasses changes secondary to cell death by any mechanism, including apoptosis (MAJNO and JORIS 1995). Cell death which occurs by nonapoptotic mechanisms is generally characterized by increased membrane permeability and cell swelling, vacuolization, blebbing of the membrane, random degradation of chromatin, and eventually cell lysis.

The activation of certain **c**ysteinyl **asp**artate-specific prote**ases** (caspases), some of which have homologues in both mammals and nematodes, appears to be a critical component of the apoptotic process (ALNEMRI et al. 1996). These caspases, which include the interleukin-1β-converting enzyme-like family proteases, CPP-32/apopain, and FLICE, cleave specific substrates involved in cellular homeostasis, repair, and structure, for example, poly(ADP-ribose)polymerase, an enzyme involved in DNA repair and genome integrity, and lamin A (for review see ZHIVOTOVSKY et al. 1996). The control of unactivated, procaspases is of obvious importance and may be mediated by endogenous serpins and inhibitor of apoptosis proteins. Recent studies suggest that activation of the caspase cascade may be an indirect result of permeability transitions resulting in the opening of mitochondrial megachannels (ZAMZAMI et al. 1996). The role of mitochondria in apoptosis is unclear, but several mitochondrial-associated factors have been implicated in the induction and inhibition of apoptosis. These include Bcl-2, which generally behaves as an apoptosis antagonist, and cytochrome *c*, which can induce apoptosis by mediating cleavage and activation of certain caspases (BOISE et al. 1995; LIU et al. 1996). Activation of calpains, a family of Ca^{2+}-dependent cysteine proteases, has been shown to be required for dexamethasone-induced apoptosis of thymocytes, and may play a role in apoptosis in other systems as well (SQUIER and COHEN 1997; SQUIER et al. 1994). Calpain substrates include α-fodrin, a cytoskeletal, actin-binding protein, whose cleavage may disrupt cell structure.

One of the dramatic differences between apoptotic and nonapoptotic cell death is the means by which dead cells are removed. Apoptotic cells typically remain membrane-bound, do not elicit an inflammatory response, and are phagocytosed by local macrophages or nonprofessional phagocytic cells. These phagocytic cells do not recruit immune cells to the site, nor do they, or the apoptotic cells themselves, appear to induce other cells to undergo apoptosis. The efficient recognition of apoptotic cells by phagocytic cells appears to be mediated by the externalization of phosphatidylserine during apoptosis (FADOK et al. 1992). The availability of annexin V, which specifically recognizes and binds to phosphatidylserine residues, provides another assay for detection of apoptotic cells (KOOPMAN et al. 1994). In the absence of phagocytic cells, for example, cultures of cells in vitro, apoptotic cells and bodies undergo secondary necrosis and lysis. Cells that die by nonapoptotic mechanisms lyse and elicit an inflammatory response. Macrophages recruited to remove dead cells and cellular material may exacerbate tissue damage by secreting cytokines which amplify the immune response.

4 Reovirus Induces Apoptosis in Murine Fibroblast Cells

Reovirus-induced cytopathicity in cultured cells and in infected mice has been well documented. Descriptions of the changes in morphology and other cellular functions in reovirus-infected cells suggested to us that reovirus induces apoptosis. Morphological and biochemical assays for apoptosis were used to determine whether reovirus induces apoptosis in L929 (L) cells, a murine fibroblast cell line (TYLER et al. 1995).

L cells infected with reovirus strain T3D were examined by electron microscopy 18h postinfection (hpi) for morphological changes typical of apoptosis. Virus-, but not mock-infected cells showed chromatin condensation and margination, budding of the plasma membrane, cytoplasmic shrinkage, and fission into apoptotic bodies (Fig. 1). These observations indicate that infection of L cells with T3D induces apoptosis. L cells were also analyzed for the presence of oligonucleosomal-sized DNA fragments (multiples of 180–200bp), a biochemical change observed in many apoptotic cells. DNA was isolated from cells which were mock-infected or infected with reovirus strain T3D or T1L at 24 and 48hpi. The DNA was separated by agarose gel electrophoresis and analyzed by Southern blot hybridization with a ^{32}P-labeled, nick-translated L cell genomic DNA probe. Both T1L- and T3D-infected cells contained oligonucleosome-length ladders at 48, but not at 24hpi (Fig. 2). Laddering was not observed with DNA isolated from mock-infected cells at either timepoint, indicating that both T1L and T3D induce apoptosis of infected L cells.

4.1 Reovirus-Induced Apoptosis Is Strain-Specific and Determined by the S1 Gene

An acridine orange/ethidium bromide staining (AO) assay was used to quantitate the number of apoptotic cells in T3D- and T1L-infected L cell cultures. Nucleic acid was stained with the AO mixture and visualized by fluorescent microscopy. Cells which exhibited condensed chromatin and cytoplasmic shrinkage were counted as apoptotic. Neither T3D- nor T1L-infected cell cultures contained significant numbers of apoptotic cells at 24hpi, but at 48hpi T3D-infected cultures contained many more apoptotic cells than was observed with T1L (Fig. 3a). Both T3D- and T1L-infected cultures contained more apoptotic cells than mock-infected cultures at 24 and 48hpi. A similar finding was obtained when an assay for quantitating low molecular weight, fragmented DNA as an indication of apoptosis was used (Fig. 3b). The strain-specific difference in induction of apoptosis was not due to a difference in growth rates or yields of T1L and T3D since these strains grew equally well in L cells (TYLER et al. 1995). These results show that reovirus

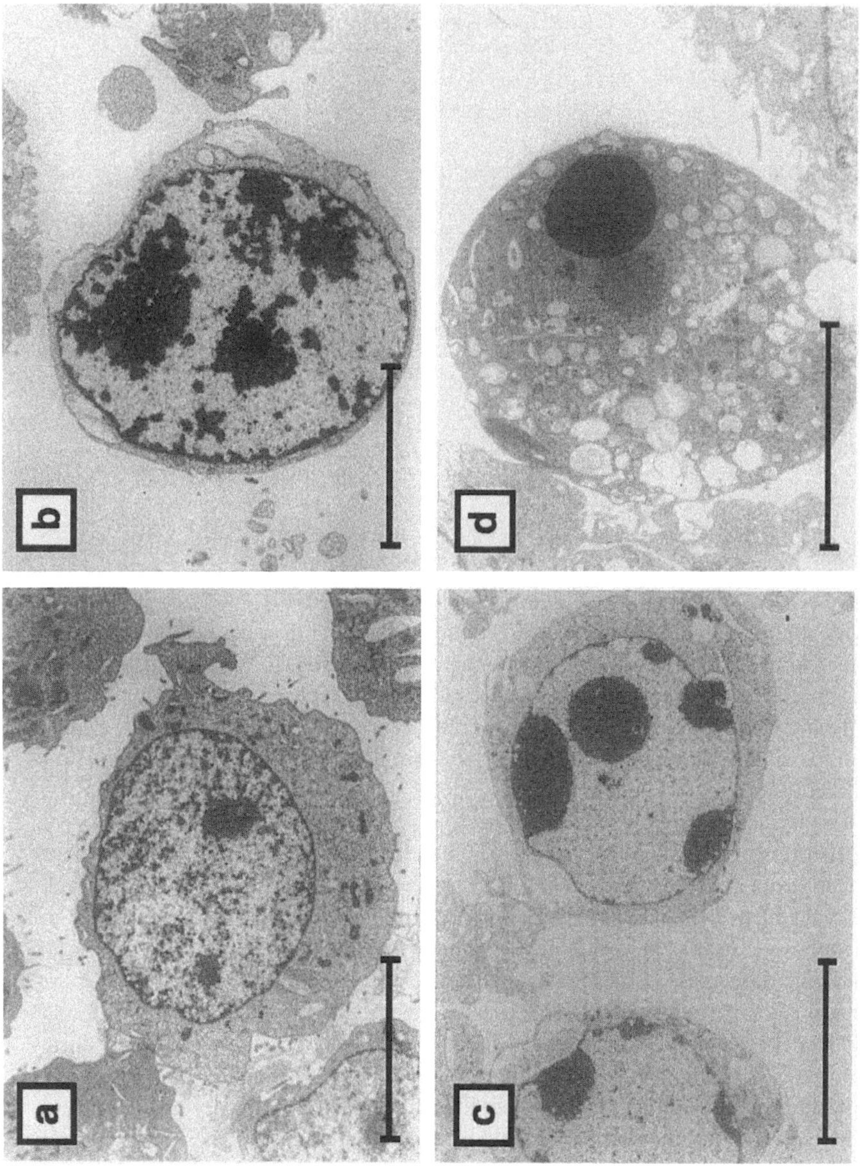

Fig. 1. Electron microscopic appearance of uninfected (**a**) and T3D-infected (**b–d**) L cells. Infected cells undergo various stages of chromatin condensation: initial condensation (**b**), margination of chromatin at the nuclear membrane (**c**), and complete condensation of the nucleus (**d**). *Bars*, 5µm. (With permission from TYLER et al. 1995)

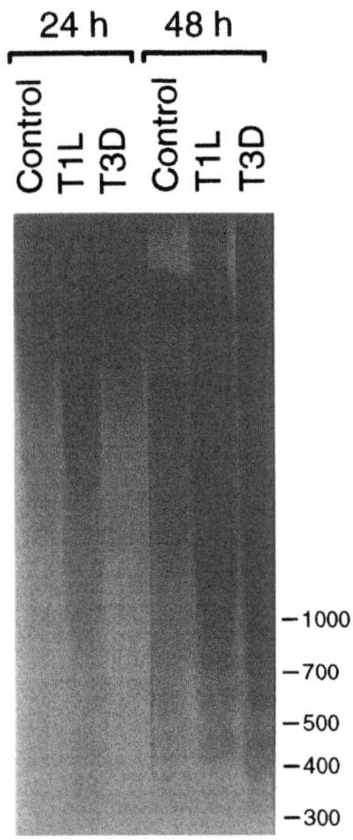

Fig. 2. Autoradiograph of Southern blot hybridization of total cellular DNA extracted from L cells which were either mock-infected or infected with T1L or T3D. At 24 or 48hpi, purified cellular DNA was resolved by agarose gel electrophoresis, transferred to a nylon membrane, and probed with [^{32}P]-labeled, nick-translated, L cell genomic DNA. *Right*, size markers (in base pairs). (With permission from TYLER et al. 1995)

induces apoptosis in a strain-specific manner, i.e., T3D induces apoptosis to a greater extent than T1L, which is independent of viral growth. We have also shown, using the AO assay, that another reovirus T3 strain, type 3 Abney (T3A) induces apoptosis to an even greater extent than T3D (TYLER et al. 1996), and that the third prototype strain, T2J induces apoptosis to a similar extent as the T3 strains (unpublished data).

The segmented nature of the reovirus double-stranded RNA genome allows for the generation of reassortant viruses when cells are coinfected with two or more reovirus strains (NIBERT et al. 1996; RAMIG and WARD 1991). The origin of each gene segment present in the reassortant progeny is determined by comparison of electrophoretic profiles of reassortant genomes with those of the parental genomes. Phenotypic differences between the parental strains can be correlated with specific gene segments by testing the reassortants for the phenotype(s) of interest. We tested 38 T1L × T3D reassortant viruses using the AO assay to determine which viral genes were associated with the different capacities of T1L and T3D to induce apoptosis. A significant association was found between the capacity of reassortant viruses to induce apoptosis and the T3D S1 gene segment and a less significant, but notable association between induction of apoptosis and the T3D M2 gene. None of the other gene segments were found to be significantly associated with differences in apoptosis induction as measured in the AO assay. The same finding was obtained when these reassortants were tested using a DNA fragmentation assay for apoptosis. These results indicate that the difference in the capacities of T1L and T3D to

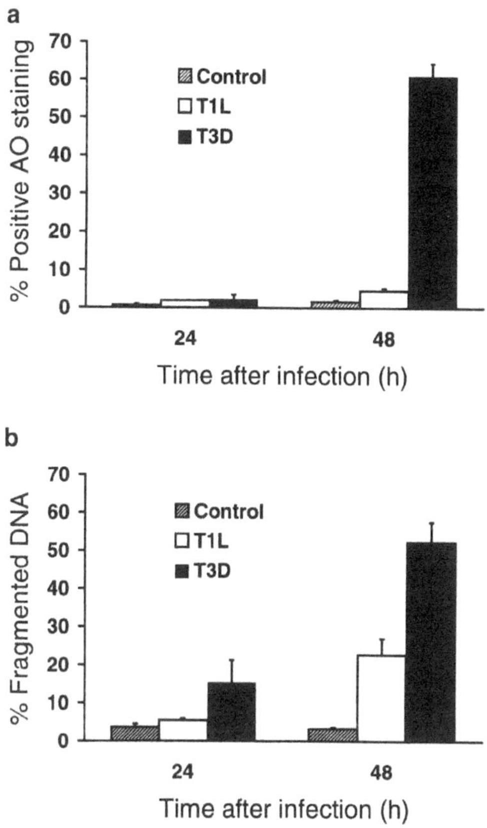

Fig. 3. Percentage of L cells undergoing apoptosis as detected by either AO staining (**a**) or fragmentation of [^{3}H]thymidine-labeled DNA (**b**). L cells were either mock-infected or infected with T1L or T3D and harvested at 24 or 48hpi for AO staining or DNA fragmentation assays. The results are expressed as the means for three independent experiments; *error bars* indicate standard deviations. (With permission from Tyler et al. 1995)

induce apoptosis is determined primarily by the S1 gene and also, but to a lesser extent, by the M2 gene.

The same analysis was carried out using a panel of 15 T1L × T3A reassortant viruses and the AO assay. As with the T1L × T3D reassortants, the S1 gene segment was found to be the primary determinant of the capacity to induce apoptosis with a small, but significant contribution from the M2 gene (Tyler et al. 1996).

The capacity of reovirus strains T3D and T1L to induce apoptosis was also analyzed using Madin-Darby canine kidney (MDCK) cells, polarized cells of epithelial origin (Rodgers et al. 1997). Both T1L and T3D were found to induce the morphological and biochemical changes indicative of apoptosis as was observed in L cells, and again, T3D did so to a greater extent than T1L. Analysis of the capacities of 25 T1L × T3D reassortant viruses to induce apoptosis in MDCK cells using the AO assay identified the S1 gene segment as the primary determinant of differences in this capacity, with a modest and independent contribution by the M2 gene. The demonstration that reovirus induces apoptosis in a strain-specific manner dictated primarily by the S1 gene in L cells and MDCK cells suggests that reovirus-induced apoptosis occurs by the same or similar mechanisms regardless of cell type.

Both the S1 and M2 gene segments encode proteins involved in early events in the viral replication cycle (McCRAE and JOKLIK 1978; MUSTOE et al. 1978). The S1 gene product, σ1 is the cell attachment protein and is an important determinant of cell tropism and virulence (LEE et al. 1981; WEINER et al. 1980a). M2 encodes the viral outer capsid protein μ1/μ1C, which, after proteolytic processing during viral disassembly, is postulated to mediate movement of virions from endosomes to the cytosol. These two genes are also involved in reovirus-mediated inhibition of host cell DNA synthesis, a phenomenon which is correlated closely with reovirus induction of apoptosis (TYLER et al. 1995). The striking linkage of the S1 and M2 genes with strain-specific differences in induction of apoptosis and inhibition of DNA synthesis suggests strongly that viral binding and entry play key roles in triggering these phenomena.

4.2 Viral Yield is not Correlated with Reovirus Induction of Apoptosis and Strain-Specific Differences in Viral Growth are Determined by Different Viral Genes

Reovirus strain T3D induces apoptosis to a much greater extent than T1L in L cells, but the two strains grow equally well in these cells, indicating that induction of apoptosis is not correlated with viral growth. The same strain-specific difference in the capacity to induce apoptosis is observed in MDCK cells, but T3D and T1L do not grow equally well in this cell type. Growth of T3D and T1L was determined by infecting monolayers of MDCK cells and incubating the cells for various time intervals (RODGERS et al. 1997). Viral titers in infected cell lysates were determined by a standard plaque assay. A difference in the growth of T1L and T3D was observed at 12–24hpi, with T1L titers exceeding those of T3D, which did not exhibit significant growth (Fig. 4).

To identify viral genes associated with the capacity to grow in MDCK cells, 25 T1L × T3D reassortant viruses were tested for growth in these cells by titering infected cell lysates 24hpi. A significant association was found between the capacities of reassortant viruses to grow in MDCK cells and the T1L L1 gene and a

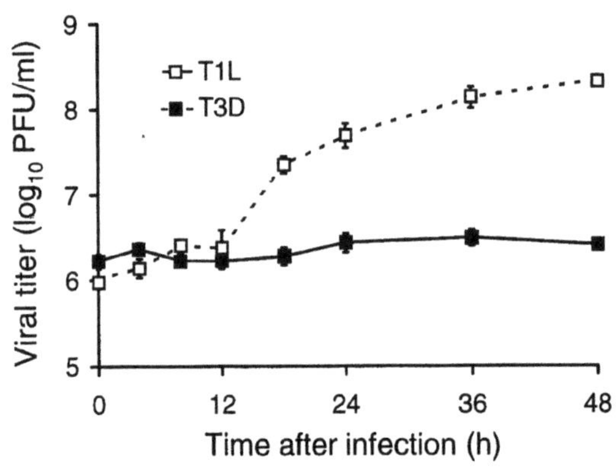

Fig. 4. Growth of T1L and T3D in MDCK cells. Monolayers of MDCK cells were infected with either T1L or T3D. After a 1h adsorption period cells were incubated for the indicated intervals, and virus in cell lysates was titrated on L cell monolayers by plaque assay. The results are presented as the means of the viral titers from three independent experiments; *error bars* indicate standard deviations of the means. (With permission from RODGERS et al. 1997)

modest association with the T1L M1 gene. No other gene segments were significantly associated with differences in viral growth in these cells. Linear regression analysis indicated that the L1 gene is the principal determinant of the difference in the capacities of T1L and T3D to grow in MDCK cells, and that the M1 gene segment has a smaller, but significant influence. The L1 and M1 gene segments encode core proteins λ3 and μ2, respectively, each with a putative role in viral RNA synthesis (McCrae and Joklik 1978; Mustoe et al. 1978). These two genes have also been linked to strain-specific differences in viral growth in other cell types (primary cultures of cardiac myocytes and aortic endothelial cells) (Matoba et al. 1991, 1993).

4.3 Viral Replication is not Required for Reovirus-Induced Apoptosis

Our results indicating that reovirus binding and entry are important events in induction of apoptosis prompted us to perform the following experiment to determine whether viral replication is required for reovirus induction of apoptosis. Purified T3D and T1L virions were UV inactivated and tested for the capacity to induce apoptosis of L cells as measured by the AO assay. The same strain-specific difference in induction of apoptosis observed with replication-competent virus was also exhibited by the UV-inactivated, replication-incompetent virions (Tyler et al. 1995). This effect was dose-dependent requiring substantially larger MOIs of UV-inactivated virions to generate similar levels of apoptosis as detected with live virus (Fig. 5). However, induction of apoptosis with the high UV-inactivated MOI of 1000 was detectable at 12h postadsorption in contrast to 48hpi using live virus. These results suggested that reovirus induction of apoptosis occurs at an early stage of the viral life cycle, i.e., binding, internalization, or disassembly, and that repli-

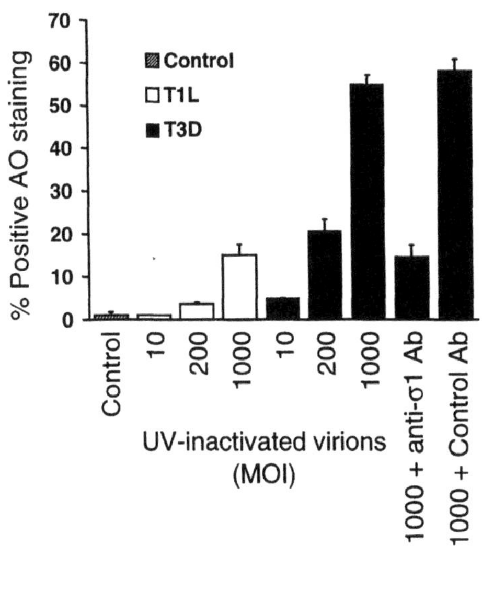

Fig. 5. Induction of apoptosis by UV-inactivated reovirus virions. Virions of T1L and T3D were inactivated by UV exposure and adsorbed to L cells at the MOIs indicated. For anti-σ1 antibody (*Ab*) and control antibody conditions, UV-inactivated T3D was preincubated with the anti-T3D σ1 MAb, 9BG5, or an isotype-matched MAb specific for T1L 1 (5C6). Cells were stained with AO 12h after adsorption of UV-inactivated virions. The results are expressed as the means for three independent experiments; *error bars* indicate standard deviations. (With permission from Tyler et al. 1995)

cation is required at low MOIs in order to generate a quantity of virus particles sufficient to induce apoptosis. This idea was tested by preincubating live T3D virus with anti-reovirus monoclonal antibodies (MAb), which vary in their capacity to block viral growth, and then inoculating L cells with the virus-MAb mixtures and quantitating apoptotic cells at 48hpi using the AO assay. Only MAbs which inhibited growth of T3D in L cells inhibited apoptosis. These results indicate that viral replication is required for induction of apoptosis at a low MOI, perhaps to generate a sufficient quantity of viral signal to induce apoptosis. At high MOIs induction of apoptosis may occur with faster kinetics because the threshold amount of viral signal required to trigger apoptosis is present without amplification through viral replication.

4.4 The Viral Cell Attachment Protein σ1, not σ1s, Determines Capacity to Induce Apoptosis

The S1 gene segment, which is the primary determinant of strain-specific differences in the capacity to induce apoptosis, encodes two proteins, σ1, a minor constituent of the virion outer capsid, and σ1s, which is found in reovirus-infected cells, but is not virion-associated (Belli and Samuel 1991; Cashdollar et al. 1989; Ceruzzi and Shatkin 1986; Ernst and Shatkin 1985; Jacobs and Samuel 1985; Sarkar et al. 1985). Preparations of purified, UV-inactivated virions do not contain σ1s protein and cannot synthesize it because they are transcriptionally inactive. The finding that purified, UV-inactivated virions induce apoptosis of L cells indicates that the S1 gene product σ1 alone determines the capacity to induce apoptosis. Preincubation of purified, UV-inactivated T3D virions with an anti-σ1 MAb significantly reduced the number of apoptotic cells, suggesting that binding of σ1 to cellular receptors triggers induction of apoptosis (Fig. 5).

5 Reovirus-Induced Apoptosis and Inhibition of DNA Synthesis is Strain-Specific and Determined by the S1 Gene

Previous studies which describe reovirus inhibition of host cell DNA synthesis share many similarities with features of reovirus-induced apoptosis. Both phenomena show the same strain-specific differences which are determined by the S1 gene, and both processes can be triggered by UV-inactivated, replication-incompetent virus (Sharpe and Fields 1981; Tyler et al. 1995). Inhibition of DNA synthesis has also been described in association with chromatin clumping and margination, which are morphological changes typical of apoptotic cells (Chaly et al. 1980). We compared the capacities of T3D, T3A, and T1L to induce apoptosis and to inhibit DNA synthesis, then tested both T1L × T3D and T1L × T3A reassortants in the same assays to determine whether reovirus-induced apoptosis and inhibition of DNA synthesis were correlated (TYLER et al. 1996).

L cells were infected with T1L or T3A, and DNA synthesis was measured by determining the amount of [^{3}H]thymidine incorporation at various intervals after adsorption. DNA synthesis inhibition was first detected with both T1L and T3A at 24hpi, but it was substantially greater with T3A, reaching a maximal difference at 48hpi (Fig. 6). To identify viral gene(s) associated with this strain-specific difference in inhibition of DNA synthesis, 20 T1L × T3A reassortants were analyzed using the same [^{3}H]thymidine incorporation assay. A significant association was found between the capacity of reassortant viruses to inhibit DNA synthesis and the S1 gene and a notable, but less significant association with the M2 gene. No other viral genes were significantly associated with strain-specific differences in DNA synthesis inhibition. Using a parametric stepwise linear regression analysis, S1 and M2 were found to contribute independently to the capacity to inhibit DNA synthesis.

We also analyzed the capacities of 37 T1L × T3D reassortants to inhibit cellular DNA synthesis and again found a significant association between this capacity and the S1 gene. Notable, but less statistically significant associations were found between strain-specific differences in DNA synthesis inhibition and the M2 gene and the L1 gene, with the M1 gene showing the least significant association. Using a parametric stepwise linear regression analysis, the S1 and M2 genes were found to contribute independently to the capacity to inhibit DNA synthesis, with S1 as the primary determinant. The M1 and L1 genes were found not to contribute significantly to the S1- and M2-determined strain-specific difference in this capacity. These results are in agreement with those from a previous study which used a small panel of T1L × T3D reassortants derived from temperature-sensitive parental viruses generated by chemical mutagenesis (SHARPE and FIELDS 1981).

The observation that T3D and T3A inhibit DNA synthesis and induce apoptosis to a much greater extent than T1L suggested that this capacity might be a serotype-dependent property. We analyzed five T1 and eight T3 field isolates for

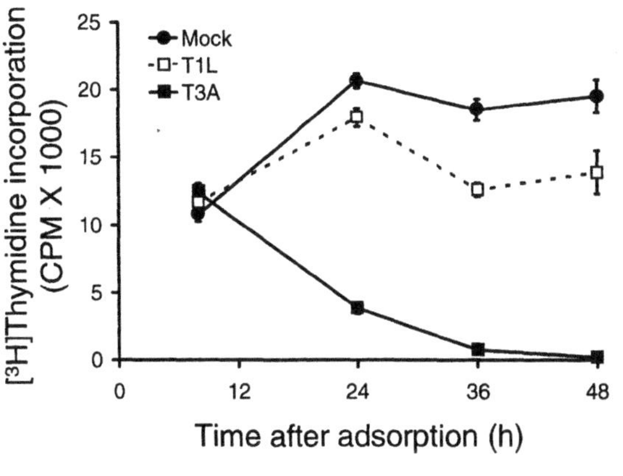

Fig. 6. Reovirus-induced inhibition of cellular DNA synthesis. L cells were either mock-infected or infected with T1L or T3A and [^{3}H] was added to the culture medium for the last 6h of the infection. Cells were harvested at the indicated times, and incorporation of [^{3}H] was measured. Results are presented as total counts per minute per well; *error bars* indicate standard error of the means. (With permission from TYLER et al. 1996)

their capacities to inhibit DNA synthesis and induce apoptosis and found that each of the T3 strains, except one, had a greater capacity to induce both phenomena than any of the T1 strains. Since viral serotype is a property determined by the S1 gene, these results are in agreement with those from the analysis of reassortant viruses and support the contention that the S1 gene is the primary determinant of the capacity of reovirus strains to induce apoptosis and inhibit DNA synthesis.

The extent to which T1L × T3A and T1L × T3D reassortant viruses induced apoptosis and inhibited DNA synthesis were compared using stepwise parametric linear regression analysis to determine the correlation between these two phenomena. Analysis of the 15 T1L × T3A reassortants and of the 37 T1L × T3D reassortants indicate a strong linear relationship between the capacities of T1L × T3A and T1L × T3D reassortants to inhibit cellular DNA synthesis and to induce apoptosis.

5.1 Reovirus-Mediated Inhibition of DNA Synthesis May Trigger Apoptosis

The correlation between inhibition of DNA synthesis and induction of apoptosis, and the primary role of the S1 gene in both processes strongly suggests that the two phenomena are linked. The nature of their association is not yet clear; however, induction of apoptosis has been shown to occur in other systems as a result of cell cycle arrest in G_0 or G_1 (Freeman and Estus 1994; Meikrantz and Schlegel 1995; Ucker 1991). Reovirus infection induces a time-dependent, reversible arrest at the G_1-S phase of the cell cycle, an event which might result from reovirus-mediated inhibition of DNA synthesis and lead to induction of apoptosis. Reovirus-induced inhibition of host cell DNA synthesis is detectable as early as 8hpi, while cellular changes indicative of apoptosis are not detected until 24–48hpi, suggesting that inhibition of DNA synthesis precedes apoptosis in infected cells.

That an early event in the viral life cycle mediates both of these processes is suggested by the following observations: (a) UV-inactivated, replication-incompetent virions can inhibit DNA synthesis and induce apoptosis (Sharpe and Fields 1981; Tyler et al. 1995), (b) the S1 gene is the primary determinant of strain-specific differences in both DNA synthesis inhibition and induction of apoptosis, and the M2 gene plays a secondary role in both of these processes (Sharpe and Fields 1981; Shaw and Cox 1973; Tyler et al. 1995, 1996), and (c) induction of apoptosis can be significantly reduced by preincubation of virus with an anti-σ1 MAb (Tyler et al. 1995). It is possible that the induction event triggers DNA synthesis inhibition and induction of apoptosis in parallel, either by the same or different signals, or one is triggered which leads to induction of the other. Reovirus-mediated inhibition of DNA synthesis and cellular proliferation can be induced by purified σ1 protein (Saragovi et al. 1995) and by an anti-reovirus T3 receptor MAb (Gaulton and Greene 1989; Saragovi et al. 1995). These data strongly suggest that DNA synthesis inhibition and by association, induction of apoptosis, are triggered by the binding of σ1 protein to cellular receptors which may transduce signals to the nucleus via kinases or second messengers. Elucidating the nature of the viral receptor and the molecular consequences of virus binding is critical to understanding the relationship between reovirus-induced inhibition of DNA synthesis and apoptosis.

6 Reovirus-Induced Apoptosis and Tissue Injury in the CNS

Reovirus infection of neonatal mice has served as a useful model system for the study of viral pathogenesis. Reoviruses infect and produce tissue injury in all major organ systems including heart, liver, lung, and the CNS (for review see TYLER and FIELDS 1996). In the brains of inoculated neonatal mice, reovirus strain T3D infects primarily neurons in the cortex, thalamus, and hippocampus, producing a lethal encephalitis. The CNS tropism and neurovirulence of reoviruses are determined by the viral S1 and M2 genes (HRDY et al. 1982; LEE et al. 1981; LUCIA-JANDRIS et al. 1993; NIBERT and FIELDS 1992; WEINER et al. 1977, 1980b), the same genes which also determine strain-specific differences in reovirus-induced DNA synthesis inhibition and apoptosis (SHARPE and FIELDS 1981; TYLER et al. 1995, 1996). Histological examination of brain tissue obtained from T3D-infected mice exhibit localized areas of neuronal destruction restricted to cortex, thalamus, and hippocampus in the absence of a significant early inflammatory response (TYLER and FIELDS 1996). These observations, and the demonstration that reovirus induces apoptosis in L cells and MDCK cells, suggested that reovirus-induced apoptosis might be an important mechanism of cell death and tissue injury in vivo.

6.1 Maximal Viral Growth is Correlated with Apoptosis and Mortality

One-day-old Swiss Webster mice were inoculated intracerebrally with 10^4 PFU of reovirus T3D diluted in gel saline or gel saline alone (mock) (OBERHAUS et al. 1997). Brain tissue was removed from euthanized animals at 1, 3, 6, and 8–9dpi (days post infection). Brain halves were assayed for viral growth in a standard plaque assay or for the presence of oligonucleosome-length DNA "ladders" indicative of apoptosis. Viral titer increased at each timepoint and reached a peak of 10^9 PFU/g at 8–9dpi (Fig. 7). Fragmentation of cellular DNA into oligonucleosome lengths was detected in tissue samples prepared from T3D-, but not mock-infected mice 8–9dpi (Fig. 7), the same timepoint at which viral titer was maximal and the mice were moribund.

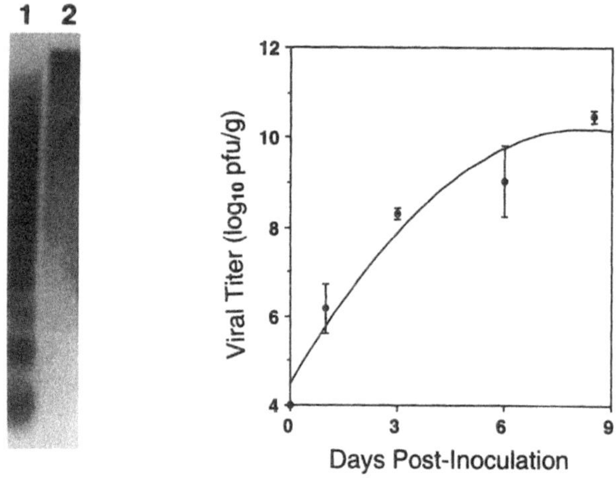

Fig. 7. a Autoradiograph of Southern blot hybridization of low molecular weight DNA extracted from T3D-(*lane 1*) or mock-(*lane 2*) infected mouse brains at 8–9 dpi. DNA was resolved by agarose gel electrophoresis, transferred to a nylon membrane and hybridized with a ^{32}P-labeled, nick-translated, L cell genomic DNA probe. **b** Growth of reovirus T3D in brain tissue. Virus titers in homogenized tissue collected at the indicated dpi were determined by plaque assay. The results are presented as the mean viral yields; *error bars* indicate standard deviations. (With permission from OBERHAUS et al. 1997)

6.2 Viral Replication, Apoptosis, and Tissue Damage are Restricted to Distinct Regions Within Infected Mouse Brain

To determine whether viral replication, apoptosis, and tissue damage occur in the same regions of infected mouse brain, tissue sections were analyzed for reovirus antigens and fragmented DNA to detect infected and apoptotic cells, respectively, at 3, 6, and 8dpi. Reovirus antigens were detected by immunocytochemistry using a rabbit polyclonal anti-T3D antiserum. Since reovirus replicates in the cytoplasm, reovirus antigen-positive cells contain a dark brown precipitate in the cytoplasm. Fragmented DNA, which is typically generated in apoptotic cells, was detected using the terminal deoxynucleotidyl transferase-mediated dUTP nick end labeling, i.e., TUNEL assay. The TUNEL procedure results in the labeling of available 3′ OH ends of DNA frequently generated during apoptosis by endonuclease-mediated cleavages. Terminal deoxynucleotidyl transferase was used to incorporate biotinylated dUTP at 3′ OH termini, which was then detected by incubation with an avidin-biotin complex and DAB staining. TUNEL positive cells contain a dark brown precipitate in the nucleus.

Only a few TUNEL- and reovirus antigen-positive cells were detected in the thalamus 3dpi, and pathological injury was minimal. At 6dpi tissue damage was pervasive within the thalamus and cingulate gyrus, and significant tissue injury was observed in cortical layer V and throughout the hippocampus, including regions

CA1 to CA3 and the dentate gyrus (Fig. 8a). Many reovirus antigen-positive and apoptotic cells were detected exclusively in these regions in the absence of inflammation (Fig. 8b–g). At 8dpi, tissue damage was even more extensive in these regions and associated with greater numbers of reovirus antigen- and TUNEL-positive cells. This is the same time point at which DNA laddering and maximal viral titer was observed and the mice were moribund. Rare TUNEL-positive cells were found in tissue sections obtained from mock-infected mice at 6 or 8dpi, but none contained reovirus antigen-positive cells or evidence of tissue damage (Fig. 9a–d).

In general, it was very difficult to distinguish individual, labeled cells so that quantitation of reovirus antigen-positive, TUNEL-positive, and doubly labeled cells was impossible. However, marked regional differences in the ratio of TUNEL- to reovirus antigen-positive signals were observed suggesting that cells located in different regions of the brain show variation in susceptibility to apoptosis. Many of the reovirus antigen-positive cells could be identified as neurons by morphological criteria since reovirus antigens present in the cytoplasm delineated axonal and dendritic processes. Neither reovirus antigen- nor TUNEL-positive cells were detected in white matter tracts, regions which do not contain neurons. These observations are consistent with the well-characterized neuronal tropism of the T3D strain.

T3D- and mock-infected tissue sections were stained with cresyl violet for analysis of morphological changes associated with apoptosis. Cells with the typical features of apoptosis, i.e., reduced cytoplasmic volume, condensed and marginated chromatin, and apparent apoptotic nuclear and cellular bodies, were observed in those regions in which reovirus antigen- and TUNEL-positive cells and tissue injury were also found in infected brain tissue, but not in tissues from mock-infected animals (Fig. 9e,f).

The colocalization and parallel temporal increase in viral antigen, apoptotic cells (identified by both morphological and biochemical characteristics), and pathological injury strongly suggest that reovirus-induced apoptosis plays a major role in the cell loss and tissue injury leading to the death of T3D-infected mice.

6.3 Apoptotic Cells in Reovirus-Infected Regions of the Brain are Rarely Reovirus Antigen-Positive

Brain sections from T3D-infected mice were analyzed using a double-labeling technique for the presence of reovirus antigens (by immunofluorescence) and apoptosis (by TUNEL) to determine whether apoptotic cells were productively infected. Many cells were labeled for either reovirus antigens or TUNEL (Fig. 10). Both reovirus antigens and fragmented DNA were detected in some cells, but it was difficult to establish a correlation between antigen staining and TUNEL with single cells when they had shrunk or fissured into apoptotic bodies. These results suggest that apoptosis in reovirus-infected brain occurs in both productively infected and

a

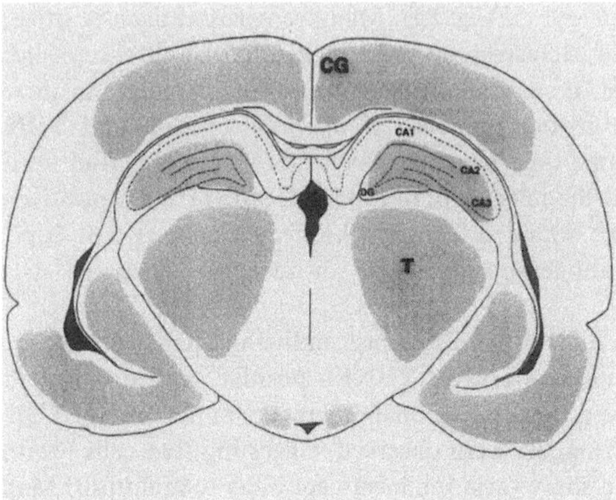

b c d e f g

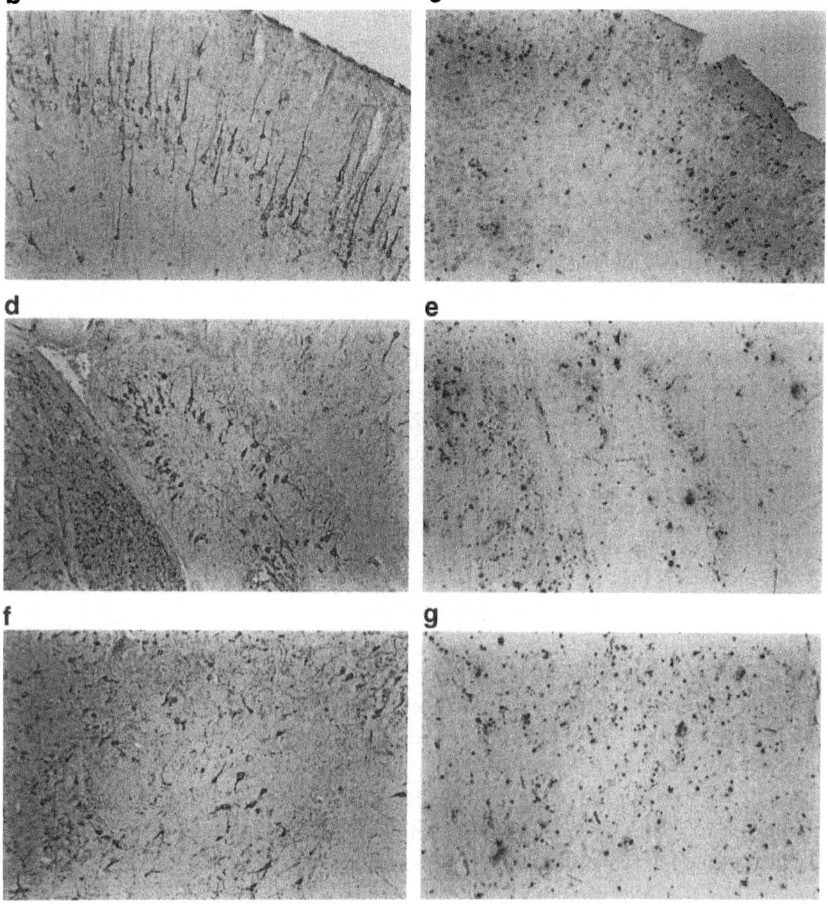

Fig. 8. a Schematic representation of a coronal section of mouse brain. *Shaded areas*, regions containing reovirus-infected and apoptotic cells, and tissue damage, 6 days after inoculation with T3D. *CG*, Cingulate gyrus; *DG*, dentate gyrus; *T*, thalamus. **b–g** Reovirus antigen positive (**b**, **d**, **f**) and TUNEL (**c**, **e**, **g**) positive cells at 6 dpi within the cortex (**b**, **c**), the dentate gyrus (**d**, **e**), and the thalamus (**f**, **g**). Positive immunostaining for reovirus antigens appears as a dark precipitate in the cytoplasm, including neuronal processes. TUNEL-positive cells contain a dark precipitate in the nucleus. Original magnification, × 25. (With permission from OBERHAUS et al. 1997)

a b

c d

e f

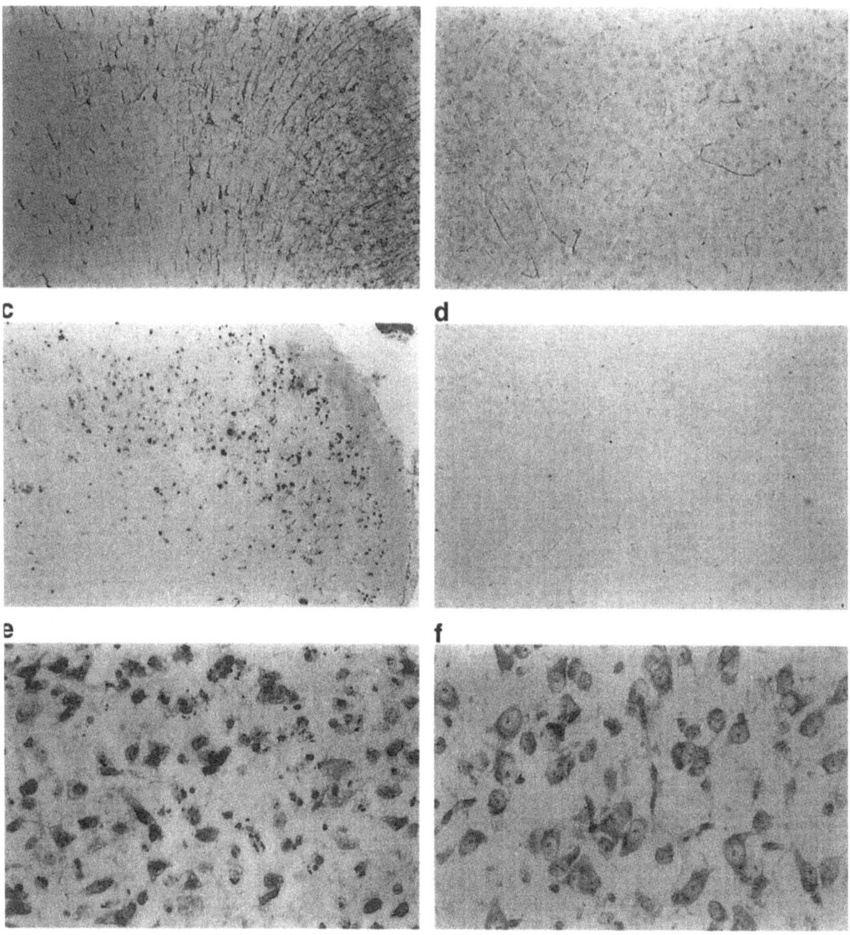

Fig. 9. Immunostaining for reovirus antigens in T3D (**a**) and mock (**b**) infected cingulate gyrus at 6 dpi; TUNEL assay on sections from T3D (**c**) and mock (**d**) infected cingulate gyrus at 6 dpi; cresyl violet staining of T3D (**e**)-and mock (**f**)-infected thalamus at 8 dpi. Original magnification, **a–d** × 25; **e**, **f** × 100. (With permission from OBERHAUS et al. 1997)

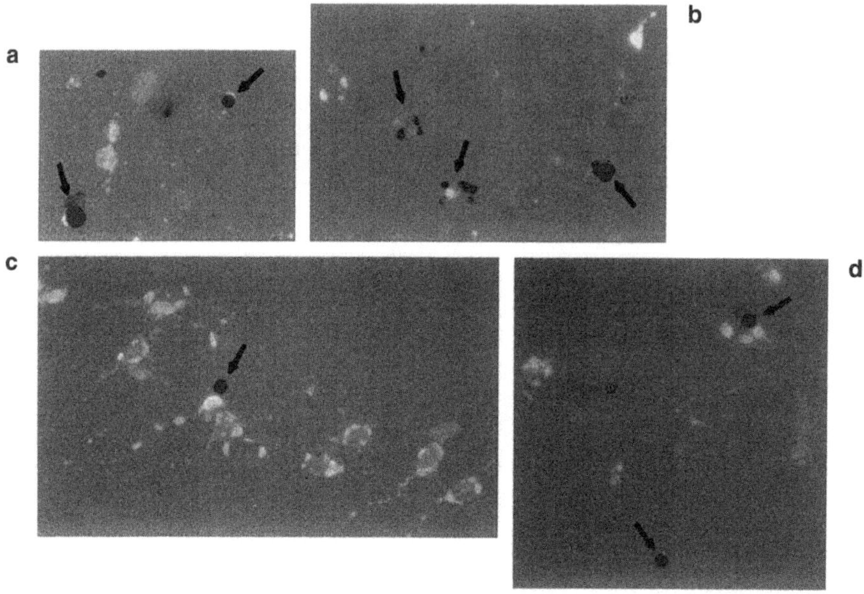

Fig. 10a–d. Double-labeling assay for reovirus antigens (FITC) and TUNEL (dark precipitate) in T3D-infected sections. A double exposure under fluorescent and light microscopy allows detection of both reovirus antigen- and TUNEL-positive cells. **a** Cells within the cingulate gyrus, 8 dpi. *Arrows*, two cells which are colabeled for reovirus antigens and fragmented DNA (TUNEL). Antigen staining appears as a thin halo surrounding the dark, TUNEL-positive nucleus as a result of cytoplasmic shrinkage. **b** Cortical cells, 8 dpi. *Arrows*, cells which appear to be colabeled for reovirus antigens and fragmented DNA and have formed multiple apoptotic bodies. **c** Cells within the hippocampus, 6 dpi. Multiple neurons contain reovirus antigens but are TUNEL-negative. *Arrow*, a TUNEL-positive nucleus adjacent to antigen-positive cytoplasm of a neuron. **d** Cells within the cingulate gyrus, 8 dpi. *Upper arrow*, a TUNEL-positive nucleus adjacent to reovirus antigen-positive bodies; *lower arrow*, a TUNEL-positive nucleus in the absence of reovirus antigen staining. Original magnification, × 100. (With permission from Oberhaus et al. 1997)

uninfected cells. It is possible that detection of infected cells by immunostaining results in an underestimation of infected, apoptotic cells due to proteolysis and/or cytoplasmic shrinkage occurring during apoptosis (MacDonald et al. 1980; Majno and Joris 1995). Protein degradation may result in altered or degraded antigenic sites, and shrinkage of the cytoplasm may preclude detection of cytoplasmic proteins in general. Cells which are infected but do not exhibit the characteristics of apoptotic cells may be at a stage prior to induction of apoptosis or may not yet have the detectable changes associated with apoptosis.

Both direct and indirect mechanisms of apoptosis induction have been reported for other viral infections. Human immunodeficiency virus and human herpesvirus 6 have both been reported to induce apoptosis predominantly in uninfected bystander cells (Finkel et al. 1995; Inoue et al. 1997) while in Sindbis virus infection of mouse brain tissue the majority of apoptotic cells appear to be infected (Lewis et al. 1996).

7 Cellular Factors Involved in Reovirus-Induced Apoptosis

The identification of cellular factors involved uniquely in virus-induced apoptosis and those which appear common to the apoptotic pathway regardless of the inducing stimulus will elucidate the events which lead to and comprise the apoptotic process as a result of virus-cell interactions.

We have already identified the reovirus σ1 protein as an important viral factor in induction of apoptosis and have begun to analyze cellular factors which have been shown to be involved in the apoptotic pathway in other systems. Bcl-2 is a 26-kDa protein detected in the membranes of mitochondria and nuclei (DE JONG et al. 1992; HOCKENBERY et al. 1990), which inhibits apoptosis induced by many but not all induction stimuli (BOISE et al. 1995; KORSMEYER 1992). Control of apoptosis appears to depend on the interaction of Bcl-2 and other members of the Bcl-2 family, for example, Bax, Bcl-X_L, Bcl-X_s, Bak, Bad, etc. (FARROW and BROWN 1996). Overexpression of Bcl-2 has been shown to inhibit apoptosis induced by several RNA viruses (HINSHAW et al. 1994; LEVINE et al. 1993).

To determine whether overexpression of Bcl-2 inhibits reovirus-induced apoptosis MDCK cells which had been genetically engineered to overexpress the human *bcl-2* gene (HINSHAW et al. 1994) were infected with T3D and analyzed for apoptosis using the AO assay for apoptotic cells (RODGERS et al. 1997). T3D-induced apoptosis was significantly reduced in MDCK cells overexpressing Bcl-2, but not in the control cells which contained the expression vector without the *bcl-2* gene (Fig. 11a). The same conclusion was reached when the cells were analyzed for oligonucleosomal DNA cleavage; DNA laddering was substantially reduced in T3D-infected cells overexpressing Bcl-2, but was clearly evident in the control cells (Fig. 11b). These results indicate that overexpression of Bcl-2 inhibits reovirus-induced apoptosis. In addition, overexpression of Bcl-2 did not significantly affect viral growth (RODGERS et al. 1997). The observation that inhibition of reovirus-induced apoptosis by Bcl-2 is not linked to viral growth is consistent with our other studies indicating that induction of apoptosis occurs independently of viral growth (TYLER et al. 1995).

p53 is a transcription factor involved in cellular DNA repair which can suppress cell growth by induction of G_1 arrest or induce apoptosis of cells which have sustained irreparable genetic damage (YONISH-ROUACH 1996). Since reovirus infection has been shown to inhibit DNA synthesis and block cell progression to S phase, we have carried out preliminary experiments to determine whether reovirus-induced apoptosis is dependent on expression of p53. Primary fibroblast cells, isolated from wild-type and p53 knockout mice, were infected with T3D and analyzed for apoptosis using the AO assay. Our preliminary results indicate that T3D-induced apoptosis is unaffected by p53 expression, suggesting that reovirus induction of apoptosis occurs by a p53-independent pathway (unpublished data).

Inhibitors of calpain, a Ca^{2+}-dependent cysteine protease, have been shown to block dexamethasone-induced apoptosis of thymocytes, suggesting a critical role for calpain in the apoptotic pathway in lymphoid cells (SQUIER and COHEN 1997).

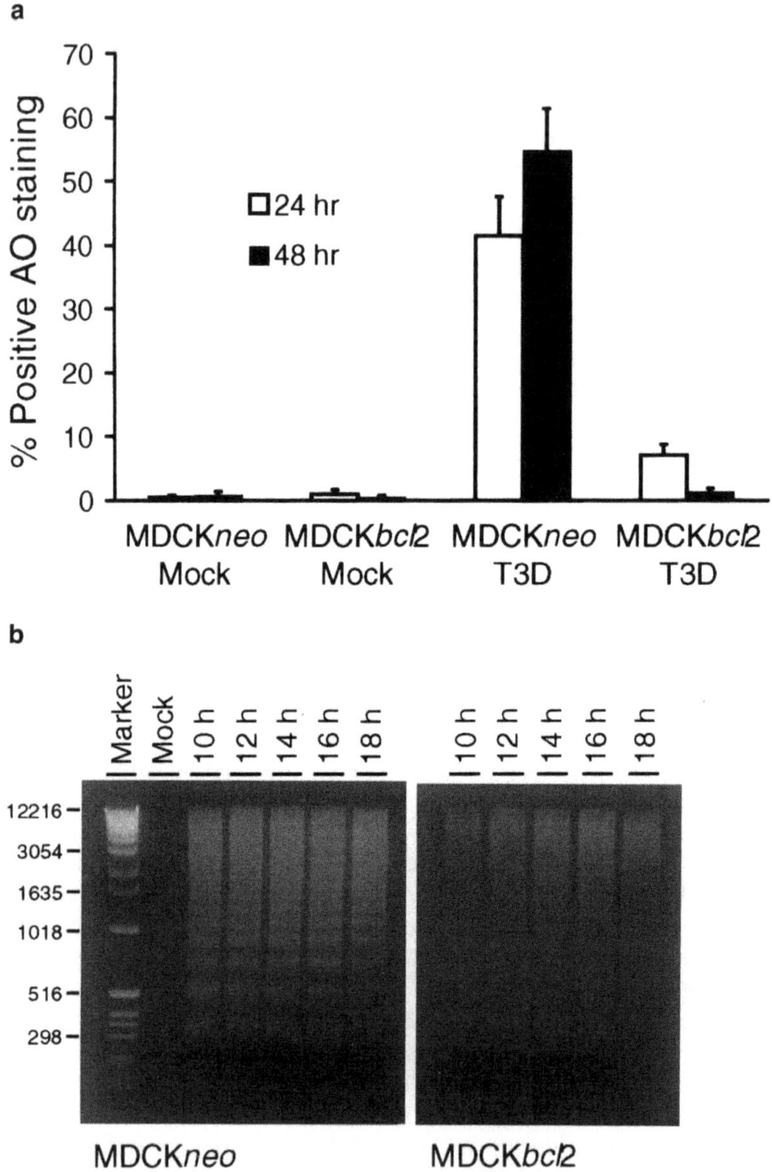

Fig. 11. **a** Percentage of MDCK*neo* and MDCK*bcl2* cells infected with T3D undergoing apoptosis as detected by the AO staining assay. Cells were either mock-infected (*control*) or infected with T3D, incubated for the indicated intervals, and harvested for AO staining. The results are expressed as the means of the data obtained in three independent experiments; *error bars* indicate standard deviations of the means. **b** Agarose gel electrophoresis of total cellular DNA extracted from MDCK*neo* and MDCK*bcl2* cells infected with T3D. Cells were either mock-infected (*control*) or infected with T3D and incubated for the indicated intervals. Cellular DNA was isolated, resolved by agarose gel electrophoresis and stained with ethidium bromide. Size markers (in base pairs) are indicated on the left. (With permission from Rodgers et al. 1997)

We have conducted preliminary studies of the effects of several calpain inhibitors on reovirus-induced apoptosis in murine L cells. All of the calpain inhibitors tested thus far inhibit T3D-induced apoptosis in L cells without significant effects on viral growth (unpublished data). These results suggest that calpains also play a role in viral-induced apoptosis.

8 Reovirus as a Model System for the Study of Apoptosis

Virus-induced apoptosis has been postulated to represent a cellular response to infection that acts to limit virus replication and spread by eliminating infected cells before completion of the viral life cycle. Our studies of reovirus-induced apoptosis and viral growth both in vitro and in vivo indicate that this putative defense mechanism is ineffective in limiting reovirus replication and cytopathicity. Many viruses which have been shown to induce apoptosis also have the capacity to inhibit or delay apoptosis (SHEN and SHENK 1995). We have identified the viral protein, σ1, as a primary determinant in the capacity to induce apoptosis, but have not determined whether reovirus also encodes an inhibitor of apoptosis. Reovirus infection of many cell types results in cell death, but reovirus can establish persistent infections under some circumstances (DERMODY et al. 1993). Viruses isolated from persistently infected cultures (termed PI viruses) have been shown to contain mutations in the genes encoding the σ1 (S1) and σ3 (S4) proteins (WETZEL et al. 1997; WILSON et al. 1996). In addition to mutant viruses, mutant cells are selected during persistent infection which do not support efficient viral entry (DERMODY et al. 1993). Mutant viruses and cells selected during persistent infection provide an interesting counterpart to cytolytic reoviruses and parental cells for further studies of the interactions between viruses and their host cells which do or do not lead to apoptosis. A similar issue arises from the characteristic age-dependent susceptibility of mice to encephalitis caused by T3 reovirus infection. Inoculation of newborn mice, before 8–10 days of age, with T3 reovirus strains results in a fatal encephalitis, but inoculation with the same viruses after 8–10 days of age are nonlethal (TARDIEU et al. 1983). It is interesting to speculate that developmentally related alterations in cellular factors or their interaction with virus may influence virus-induced apoptosis or cell susceptibility to induction of apoptosis.

Strain-specific differences in reovirus-induced apoptosis are determined primarily by the S1 gene and to a lesser extent by the M2 gene (TYLER et al. 1995). These genes are also important in tropism and virulence in vivo. The T3 S1 gene segment is associated with maximal induction of apoptosis in L cells and MDCK cells, and T3D infection of mouse brain is associated with substantial apoptosis and tissue injury in specific regions for which T3D has a predilection. T1L also induces apoptosis in L and MDCK cells, but to a much lesser extent than T3D. However, T1L specifically infects and is associated with cytopathicity of ependymal cells in infected mouse brain. It remains to be determined whether the strain-specific difference in the capacity to induce apoptosis in vitro will be observed in vivo, and whether reovirus infection and tissue injury observed in organs other than the brain are also associated with apoptosis.

The mechanism by which viruses induce and sometimes alter the course of apoptosis is likely to vary with the virus and cell type and to involve a complex interplay among viral and cellular factors. The demonstration that reovirus-induced apoptosis is correlated with cytopathic effects in cultured cells and pathological tissue injury in the brains of infected mice provides a unique and well-defined model system for studying the mechanisms by which viruses induce apoptosis and the role of this process in viral pathogenesis.

References

Alnemri ES, Livingston DJ, Nicholson DW, Salvesen G, Thornberry NA, Wong WW, Yuan J (1996) Human ICE/CED-3 protease nomenclature. Cell 87:171

Belli BA, Samuel CE (1991) Biosynthesis of reovirus-specified polypeptides: expression of reovirus S1 encoded sigma 1NS protein in transfected and infected cells as measured with serotype specific polyclonal antibody. Virology 185:698–709

Boise LH, Gottschalk AR, Quintans J, Thompson CB (1995) Bcl-2 and Bcl-2-related proteins in apoptosis regulation. In: Kroemer G, Martinez AC (eds) Apoptosis in Immunology. Springer, Berlin Heidelberg New York, pp 107–121 (Current topics in microbiology and immunology, vol 200)

Bursch W, Oberhammer F, Schulte-Hermann R (1992) Cell death by apoptosis and its protective role against disease. Trends Pharmacol Sci 13:245–251

Cashdollar LW, Blair P, Van Dyne S (1989) Identification of the sigma 1s protein in reovirus serotype 2-infected cells with antibody prepared against a bacterial fusion protein. Virology 168:183–186

Ceruzzi M, Shatkin AJ (1986) Expression of reovirus p14 in bacteria and identification in the cytoplasm of infected mouse L cells. Virology 153:35–45

Chaly N, Johnstone M, Hand R (1980) Alterations in nuclear structure and function in reovirus-infected cells. Clin Invest Med 2:141–152

Cohen JJ, Duke RC (1984) Glucocorticoid activation of a calcium dependent endonuclease in thymocyte nuclei leads to cell death. J Immunol 132:38–42

de Jong D, Prins F, Van Krieken HHJM, Mason DY, Van Ommen GB, Kluin PM (1992) Subcellular localization of bcl-2 protein. In: Potter M, Melchers F (eds) Mechanisms in B cell neoplasia. Springer, Berlin Heidelberg New York, pp 287–292 (Current topics in microbiology and immunology, vol 176)

Dermody TS, Nibert ML, Wetzel JD, Tong X, Fields BN (1993) Cells and viruses with mutations affecting viral entry are selected during persistent infections of L cells with mammalian reoviruses. J Virol 67:2055–2063

Duke DC, Chervenak R, Cohen JJ (1983) Endogenous endonuclease-induced DNA fragmentation: an early event in cell mediated cytolysis. Proc Natl Acad Sci USA 80:6361–6365

Duncan MR, Stanish SM, Cox DC (1978) Differential sensitivity of normal and transformed human cells to reovirus infection. J Virol 28:444–449

Ensminger WD, Tamm I (1969a) Cellular DNA and protein synthesis in reovirus-infected cells. Virology 39:357–359

Ensminger WD, Tamm I (1969b) The step in cellular DNA synthesis blocked by reovirus infection. Virology 39:935–938

Ernst H, Shatkin AJ (1985) Reovirus hemagglutinin mRNA codes for two polypeptides in overlapping reading frames. Proc Natl Acad Sci USA 82:48–52

Fadok VA, Voelker DR, Campbell PA, Cohen JJ, Bratton DL, Henson PM (1992) Exposure of phosphatidyl serine on the surface of apoptotic lymphocytes triggers specific recognition and removal by macrophages. J Immunol 148:2207–2216

Farone AL, O'Brien PC, Cox DC (1993) Tumor necrosis factor-alpha induction by reovirus serotype 3. J Leukocyte Biol 53:133–137

Farrow SN, Brown R (1996) New members of the Bcl-2 family and their protein partners. Curr Opin Genet Dev 6:45–49
Finkel TH, Tudor-Williams G, Banda NK, Cotton MF, Curiel T, Monks C, Baba TW, Ruprecht RM, Kupfer A (1995) Apoptosis occurs predominately in bystander cells and not in productively infected cells of HIV- and SIV-infected lymph nodes. Nat Med 1:129–134
Freeman R, Estus SEM (1994) Analysis of cell-cycle related gene expression in postmitotic neurons: selective induction of cyclin D1 during programmed cell death. Neuron 12:343–355
Gaulton GN, Greene MI (1989) Inhibition of cellular DNA synthesis by reovirus occurs through a receptor-linked signaling pathway that is mimicked by antiidiotypic, antireceptor antibody. J Exp Med 169:197–211
Gauntt CJ (1973) Induction of interferon in L cells by reoviruses. Infection Immun 7:711–717
Gòmatos PJ, Tamm I (1963) Macromolecular synthesis in reovirus infected L cells. Biochim Biophys Acta 72:651–653
Gorman AM, McGowan A, O'Neill C, Cotter T (1996) Oxidative stress and apoptosis in neurodegeneration. J Neurol Sci 139:45–52
Hand R, Tamm I (1973) Reovirus: effect of noninfective viral components on cellular deoxyribonucleic acid synthesis. J Virol 11:223–231
Hand R, Ensminger WD, Tamm T (1971) Cellular DNA replication in infections with cytocidal RNA viruses. Virology 44:527–536
Henderson DR, Joklik WK (1978) The mechanism of interferon induction by UV-irradiated reovirus. Virology 91:389–406
Hinshaw VS, Olsen CW, Dybdahl-Sissoko N, Evans D (1994) Apoptosis: a mechanism of cell killing by influenza A and B viruses. J Virol 68:3667–3673
Hockenbery D, Nunez G, Milliman C, Schreiber RD, Korsmeyer SJ (1990) Bcl-2 is an inner mitochondrial membrane protein that blocks programmed cell death. Nature 348:334–336
Houge G, Robaye B, Elkhom TS, Golstein J, Mellgren G, Gjertsen BT, Lanotte M, Doskeland SO (1995) Fine mapping of 28S rRNA sites specifically cleaved in cells undergoing apoptosis. Mol Cell Biol 15:2051–2062
Hrdy DB, Duby DN, Fields BN (1982) Molecular basis of reovirus neurovirulence: role of the M2 gene in avirulence. Proc Natl Acad Sci USA 79:1298–1302
Inoue Y, Yasukawa M, Fujita S (1997) Induction of T-cell apoptosis by human herpesvirus 6. J Virol 71:3751–3759
Jacobs JL, Ferguson RE (1991) The Lang strain of reovirus serotype 1 and the Dearing strain of reovirus serotype 3 differ in their sensitivities to beta interferon. J Virol 65:5102–5104
Jacobs JL, Samuel CE (1985) Biosynthesis of reovirus-specified polypeptides: the reovirus S1 mRNA encodes two primary translation products. Virology 143:63–74
Kerr JFR, Wyllie AH, Currie AR (1972) Apoptosis: a basic biological phenomenon with wide-ranging implications in tissue kinetics. Br J Cancer 26:239–257
Koopman G, Reutelingsperger CPM, Kuijten GAM, Keehnen RMJ, Pals ST, Van Oers MHJ (1994) Annexin V for flow cytometric detection of phosphatidyl-serine expression on B cells undergoing apoptosis. Blood 84:1415–1420
Korsmeyer SJ (1992) Bcl-2 initiates a new category of oncogenes: regulators of cell death. Blood 80: 879–886
Lai M-HT, Joklik WK (1973) The induction of interferon by temperature-sensitive mutants of reovirus, UV-irradiated reovirus, and subviral reovirus particles. Virology 51:191–204
Lai M-HT, Werenne JJ, Joklik WK (1973) The preparation of reovirus top component and its effect on host DNA and protein synthesis. Virology 53:237–244
Lee PW, Hayes EC, Joklik WK (1981) Protein sigma 1 is the reovirus cell attachment protein. Virology 108:156–163
Levine B, Huang Q, Issacs JT, Reed JC, Griffin DE, Hardwick JM (1993) Conversion of lytic to persistent alphavirus infection by the bcl-2 oncogene. Nature 361:739–742
Lewis J, Wesselingh SL, Griffin DE, Hardwick JM (1996) Alphavirus-induced apoptosis in mouse brains correlates with neurovirulence. J Virol 70:1828–1835
Liu X, Kim CN, Yang J, Jemmerson R, Wang X (1996) Induction of apoptotic program in cell-free extracts: requirement for dATP and cytochrome c. Cell 86(1):147–157
Long WF, Burke DC (1971) Interferon production by double-stranded RNA: a comparison of interferon induction by reovirus RNA to that by a synthetic double-stranded polynucleotide. J Gen Virol 12: 1–11

Lucia-Jandris P, Hooper JW, Fields BN (1993) Reovirus M2 gene is associated with chromium release from L cells. J Virol 67:5339–5345
MacDonald RG, Martin TP, Cidlowski JA (1980) Glucocorticoids stimulate protein degradation in lymphocytes: a possible mechanism of steroid-induced cell death. Endocrinology 107:1512–1524
Majno G, Joris I (1995) Apoptosis, oncosis, and necrosis. Am J Pathol 146:3–15
Matoba Y, Sherry B, Fields BN, Smith TW (1991) Identification of the viral genes responsible for growth of reovirus in cultured mouse heart cells. J Clin Invest 87:1628–1633
Matoba Y, Colucci WS, Fields BN, Smith TW (1993) The reovirus M1 gene determines the relative capacity of growth of reovirus in cultured bovine aortic endothelial cells. J Clin Invest 92:2883–2888
McCrae MA, Joklik WK (1978) The nature of the polypeptide encoded by each of the ten double-stranded RNA segments of reovirus type 3. Virology 89:578–593
Meikrantz W, Schlegel R (1995) Apoptosis and the cell cycle. J Cell Biochem 58:160–174
Mustoe TA, Ramig RF, Sharpe AH, Fields BN (1978) Genetics of reovirus: identification of the dsRNA segments encoding the polypeptides of the mu and sigma size classes. Virology 89:594–604
Nibert ML, Fields BN (1992) A carboxy-terminal fragment of μ1/μ1c is present in infectious subvirion particles of mammalian reoviruses and is proposed to have a role in penetration. J Virol 66:6408–6418
Nibert ML, Schiff LA, Fields BN (1996) Reoviruses and their replication. In: Fields BN, Knipe DM, Howley PM et al (eds) Fields virology, 3rd edn. Raven, New York, pp 1557–1596
Oberhaus SM, Smith RL, Clayton GH, Dermody TS, Tyler KL (1997) Reovirus infection and tissue injury in the mouse central nervous system are associated with apoptosis. J Virol 71:2100–2106
Ramig RF, Ward RL (1991) Genomic segment reassortment in rotaviruses and other reoviridae. Adv Virus Res 39:163–207
Rodgers SE, Barton ES, Oberhaus SM, Pike B, Gibson CA, Tyler KL, Dermody TS (1997) Reovirus-induced apoptosis of MDCK cells is not linked to viral yield and is blocked by Bcl-2. J Virol 71:2540–2546
Saragovi HU, Bhandoola A, Lemercier MA, Mohamad Akbar GK, Greene MI (1995) A receptor that subserves reovirus binding can inhibit lymphocyte proliferation triggered by mitogenic signals. DNA Cell Biol 14:653–664
Sarkar G, Pelletier J, Bassel-Duby R, Jayasuriya A, Fields BN, Sonnenberg N (1985) Identification of a new polypeptide coded by reovirus gene S1. J Virol 54:720–725
Schwartzman RA, Cidlowski JA (1993) Apoptosis: the biochemistry and molecular biology of programmed cell death. Endocr Rev 14:133–151
Sharpe AH, Fields BN (1981) Reovirus inhibition of cellular DNA synthesis: role of the S1 gene. J Virol 38:389–392
Sharpe AH, Fields BN (1982) Reovirus inhibition of cellular RNA and protein synthesis: role of the S4 gene. Virology 122:381–391
Sharpe AH, Chen LB, Fields BN (1982) The interaction of mammalian reoviruses with the cytoskeleton of monkey kidney CV-1 cells. Virology 120:399–411
Shaw JE, Cox DC (1973) Early inhibition of cellular DNA synthesis by high multiplicities of infectious and UV-inactivated reovirus. J Virol 12:704–710
Shen Y, Shenk TE (1995) Viruses and apoptosis. Curr Opin Genet Dev 5:105–111
Squier MKT, Cohen JJ (1997) Calpain, an upstream regulator of thymocyte apoptosis. J Immunol 158:3690–3697
Squier MKT, Miller ACK, Malkinson AM, Cohen JJ (1994) Calpain activation in apoptosis. J Cell Physiol 159:229–237
Tardieu M, Powers ML, Weiner HL (1983) Age dependent susceptibility to reovirus type 3 encephalitis: role of viral and host factors. Ann Neurol 13:602–607
Thompson CB (1995) Apoptosis in the pathogenesis and treatment of disease. Science 267:1456–1462
Tyler KL, Fields BN (1996) Reoviruses. In: Fields BN, Knipe DM, Howley PM et al (eds) Fields virology, 3rd edn. Raven, New York, pp 1597–1624
Tyler KL, Squier MKT, Rodgers SE, Schneider BE, Oberhaus SM, Grdina TA, Cohen JJ, Dermody TS (1995) Differences in the capacity of reovirus strains to induce apoptosis are determined by the viral attachment protein σ1. J Virol 69:6972–6979
Tyler KL, Squier MKT, Brown AL, Pike B, Willis D, Oberhaus SM, Dermody TS, Cohen JJ (1996) Linkage between reovirus-induced apoptosis and inhibition of cellular DNA synthesis: role of the S1 and M2 genes. J Virol 70:7984–7991
Ucker DS (1991) Death by suicide: one way to go in mammalian cellular development. New Biol 3:103–109

Weiner HL, Drayna D, Averill DR, Fields B (1977) Molecular basis of reovirus virulence: role of the S1 gene. Proc Natl Acad Sci USA 74:5744–5748
Weiner HL, Ault KA, Fields BN (1980a) Interaction of reovirus with cell surface receptors. I. Murine and human lymphocytes have a receptor for the hemagglutinin of reovirus type 3. J Immunol 124:2143–2148
Weiner HL, Powers ML, Fields BN (1980b) Absolute linkage of virulence and central nervous system tropism of reoviruses to viral hemagglutinin. J Infect Dis 141:609–616
Wetzel JD, Wilson GJ, Baer GS, Dunnigan LR, Wright JP, Tang DSH, Dermody TS (1997) Reovirus variants selected during persistent infections of L cells contain mutations in the viral S1 and S4 genes and are altered in viral disassembly. J Virol 71:1362–1369
Wilson GJ, Wetzel JD, Puryear W, Bassel-Duby R, Dermody TS (1996) Persistent reovirus infections of L cells select mutations in viral attachment protein σ1 that alter oligomer stability. J Virol 70:6598–6606
Wyllie AH, Morris RG, Smith AL, Dunlop D (1984) Chromatin cleavage in apoptosis: association with condensed chromatin morphology and dependence on macromolecular synthesis. J Pathol 142:67–77
Yonish-Rouach E (1996) The p53 tumour suppressor gene: a mediator of a G1 growth arrest and of apoptosis. Experientia 52:1001–1007
Zamzami N, Susin SA, Marchetti P, Hirsh T, Gomez-Monterrey I, Castedo M, Guido K (1996) Mitochondrial control of nuclear apoptosis. J Exp Med 183:1533–1544
Zhivotovsky B, Burgess DH, Vanags DM, Orrenius S (1996) Involvement of cellular proteolytic machinery in apoptosis. Biochem Biophys Res Commun 230:481–488

Pathogenesis of Reovirus Myocarditis

B. SHERRY

1 Introduction

Acute myocarditis (ARETZ et al. 1986) is prevalent in humans, with reports suggesting that 5%–20% of the population has suffered some form of viral myocarditis (BANDT et al. 1979; OKUNI et al. 1975; WOODRUFF 1980). It is often fatal in infants (CHERRY 1995; KAPLAN et al. 1983; MARTIN et al. 1994). In older individuals the acute disease usually resolves but can progress to chronic myocarditis and/or dilated cardiomyopathy (ARCHARD et al. 1991; KANDOLPH et al. 1991; KEELING et al. 1994; MARTINO et al. 1994; MATSUMORI and KAWAI 1982; MORIMOTO et al. 1992; OLSEN 1992; SOLE and LIU 1994) with concomitant cardiac failure (BORGGREFE

College of Veterinary Medicine, North Carolina State University, 4700 Hillsborough St., Raleigh, NC 27606, USA et al. 1994; Di Lenarda et al. 1994). Enteroviruses (mainly coxsackieviruses) most likely account for 20%–50% or more of the cases (Archard et al. 1991; Bowles et al. 1986; Easton and Eglin 1988; Kandolph et al. 1987, 1991; Tracy et al. 1990a,b; Weiss et al. 1991), and there is compelling evidence that enterovirus chronic myocarditis and acute myocarditis are immune-mediated (Cook et al. 1995; Guthrie et al. 1984; Hashimoto et al. 1983; Huber 1992; Huber et al. 1988; Kishimoto and Abelmann 1990; Klingel et al. 1996; Leslie et al. 1989; Liu et al. 1995; Martino et al. 1995; Rose et al. 1988, 1992; Schwimmbeck et al. 1996; Wolfgram et al. 1985; Woodruff 1980).

The virus, however, is also directly cytopathogenic (Chow et al. 1992; Grun et al. 1988; Herzum et al. 1994; Hufnagel et al. 1995), and recent evidence suggests a protective role for immune cell-generated nitric oxide (Hiraoka et al. 1996; Lowenstein et al. 1996; Mikami et al. 1996). Immune-mediated mechanisms of myocarditis have been studied extensively; however, a large variety of viruses have been implicated in the 50% or more of cases that are not due to enteroviruses (Leslie et al. 1989; Marboe and Fenoglio 1988; Matsumori et al. 1995; See and Tilles 1991; Wenger et al. 1990), and the role of immune-mediated damage in those cases is unclear. Importantly, nonimmune mediated mechanisms of viral myocarditis remain largely unexplored.

In mice reovirus-induced myocarditis presents as a startlingly different disease from enterovirus myocarditis. Reovirus myocarditis is characterized by extensive cardiac tissue necrosis with little inflammatory infiltrate (Goller et al. 1986; Hassan et al. 1965; Sherry et al. 1989; Stangl et al. 1987; Walters et al. 1965) in stark contrast to enterovirus myocarditis (Rabin et al. 1964). As described below, reovirus-induced myocarditis is not immune mediated. Moreover, infection of primary cardiac myocyte cultures with myocarditic and nonmyocarditic reoviruses has demonstrated that viral spread and cumulative myocyte death are correlated with viral myocarditic potential. Thus reovirus provides an excellent opportunity to investigate nonimmune mediated myocarditis. Reoviruses have been implicated periodically in fatal infant myocarditis, and the ubiquity of reoviruses in humans (> 70% of 4-year-olds have seroconverted; Lerner et al. 1947) presents significant opportunities for disease. Their well-characterized pathogenesis in the mouse (Tyler and Fields 1996; Virgin et al. 1997) provides an excellent model for investigating disease.

Investigations of reovirus functions in vitro and in vivo have benefited enormously from the use of classic genetic analyses, taking advantage of the segmented reovirus genome. The reovirus genome is composed of ten gene segments of double stranded RNA (dsRNA), grouped into the S, M, and L size classes (with 4, 3, and 3 segments, respectively, ranging from 1 to 4kbp each; reviewed in Nibert et al. 1996). Each gene segment encodes one protein, with the exception of segment S1. "Reassortant" viruses, containing a mixture of genes derived from two parent viruses, have been extremely powerful tools for the identification of viral genes that determine viral physical properties, replicative functions, and pathogenesis in the mouse (reviewed in Coen and Ramig 1996; Nibert et al. 1996). We have used this approach to identify viral genes that determine acute myocarditis and to compare infection parameters that vary in a large panel of closely related myocarditic and nonmyocarditic viruses. In this way reovirus genetics has provided the means to characterize the relationship between viral gene function, parameters of infection, and induction of myocarditis.

2 Reovirus-Induced Myocarditis is not Determined by Immune Cell Function

Initially we compared the gross pathogenesis of myocarditic and nonmyocarditic reovirus strains in the mouse (SHERRY et al. 1989). Histologically the acute disease differed profoundly from enterovirus-induced myocarditis in that there was marked necrosis with little inflammatory infiltrate (Fig. 1A). Electron microscopy revealed virus in dying cardiac myocytes characterized by dystrophic calcification and grossly abnormal mitochondria, suggesting a direct cytopathic effect. While reovirus can induce apoptosis in some cells (OBERHAUS et al. 1997; RODGERS et al. 1997; TYLER et al. 1995, 1996), there was no evidence of apoptosis in these cardiac cells, perhaps reflecting their differentiated state (WANG and WALSH 1996). Immunocytochemistry revealed that necrotic foci contained viral antigen, indicating that tissue damage was directly associated with viral replication (Fig. 1B).

Both acute and chronic enterovirus-induced myocarditis involve immune-mediated damage, although the virus is also cytopathogenic (see Sect. 1). We asked whether immune cells are required for, or protect against, reovirus-induced myocarditis (SHERRY et al. 1993). Introduction of immune $CD4^+$ or $CD8^+$ T cells or neutralizing antibody protected against challenge with a potently myocarditic reovirus strain, and myocarditic reovirus strains induced myocarditis in nude mice (SHERRY et al. 1989) and severe combined immunodeficiency (SCID) mice (SHERRY et al. 1993). Moreover, myocarditic reovirus strains induced myocarditis in mice depleted of macrophage-inflammatory protein-1α; (B. SHERRY and D. COOK, unpublished results), in contrast to results seen for a myocarditic coxsackievirus strain in the same mice (COOK 1996; COOK et al. 1995). In addition, nonmyocarditic reovirus strains did not induce cardiac lesions in SCID mice (SHERRY et al. 1993); thus these viruses do not have a "latent" myocarditic potential controlled by the immune response. Together the data demonstrated that reovirus-induced myocarditis is not immune-mediated and confirm that this heart disease is strikingly different from enterovirus-induced myocarditis.

3 Reovirus-Induced Myocarditis is Determined by Viral Core Proteins Involved in RNA Synthesis

To identify the viral genes that determine myocarditic potential, we took advantage of the reovirus segmented genome and used classic genetic analyses. First, we

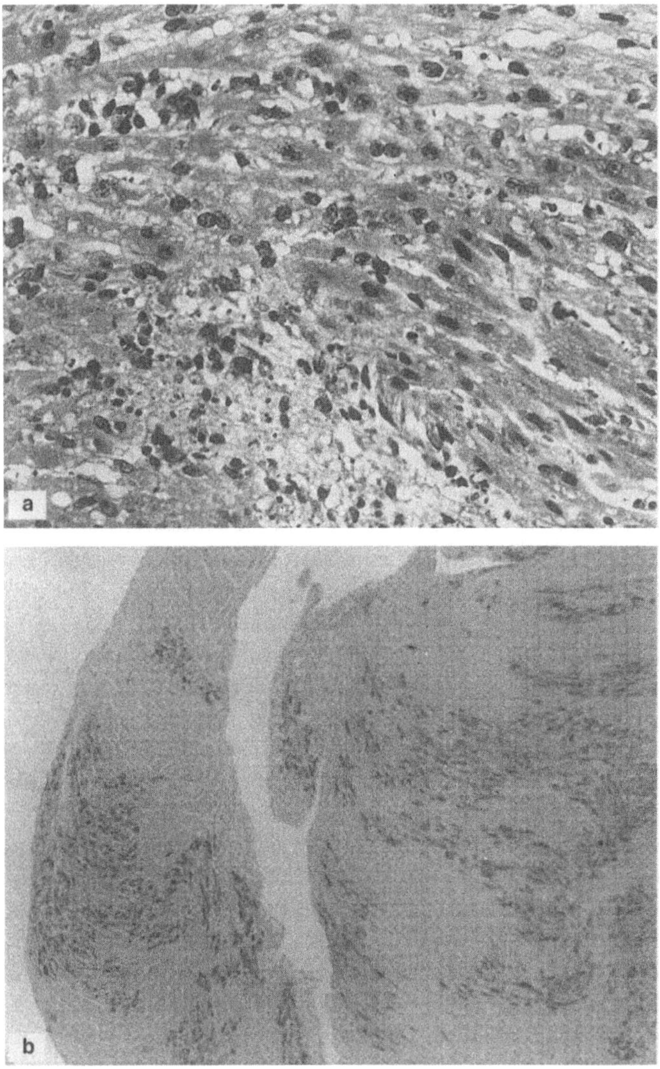

Fig. 1a,b. Cardiac sections from 8B-injected mice. Neonatal Cr:(NIH)S mice were injected intramuscularly with 10^6 PFU of the myocarditic reovirus 8B. Five days after injection mice were killed, and hearts were fixed and sectioned for H&E staining (**a**; adapted from SHERRY et al. 1989) or immunocytochemistry using hyperimmune rabbit anti-reovirus antiserum (b). Sections stained with preimmune serum were antigen-negative (data not shown)

focused on an unusually potent myocarditic reovirus reassortant, 8B (SHERRY and FIELDS 1989; SHERRY et al. 1989), which was derived from a mouse infected with a mixture of the two prototype strains serotype 3 Dearing (T3D) and serotype 1 Lang (T1L) (WENSKE et al. 1985). While T3D is completely nonmyocarditic, and T1L is poorly myocarditic, 8B is efficiently myocarditic, inducing myocarditis in all mice even when injected at low doses. Our genetic analyses (Fig. 2) using 8B-derived reassortant viruses identified the 8B M1 gene as a determinant of myocarditis ($p = 0.002$; SHERRY and FIELDS 1989). The 8B L2 gene was also implicated ($p = 0.05$; SHERRY and FIELDS 1989), as was the 8B L1 gene ($p = 0.019$; SHERRY and BLUM 1994). The 8B M1, L2, and L1 genes were all derived from the T1L parent virus (SHERRY et al. 1989), indicating that the 8B myocarditic phenotype does not reflect a novel T1L/T3D T1L gene constellation, and that instead 8B had suffered a mutation(s) relative to T1L, as yet unidentified.

To determine whether the M1, L1, and L2 genes were associated with the myocarditic potential of other reoviruses, we used reassortant viruses derived from two other myocarditic viruses for analyses (SHERRY and BLUM 1994). The M1 gene was identified as a determinant of myocarditis in both analyses, consistent with our initial observation using 8B-derived viruses. The L1 and L2 genes were associated with disease in one analysis each, again consistent with our results using 8B-derived viruses. One analysis identified the S1 gene as a determinant of myocarditis; however, its association with reovirus induction of myocarditis in general remains unclear.

Together our three genetic analyses identified the M1, L1, and L2 genes as determinants of reovirus-induced myocarditis. Interestingly, these same genes have been implicated in virulence and replication in the liver, determining hepatitis (HALLER et al. 1995). These three genes encode viral core proteins forming a structural unit at each vertex of the core (Fig. 3) (reviewed in NIBERT et al. 1996 and in this book). The L2-encoded λ2 protein forms pentameric channels (DRYDEN et al. 1993; METCALF et al. 1991) for extrusion of newly synthesized RNA from the core, where first strand RNA is synthesized. Protein λ2 is also a guanylyl transferase (CLEVELAND et al. 1986; MAO and JOKLIK 1991). The L1-encoded λ3 protein lies at the base of the λ2 pentameric channel (DRYDEN et al. 1993; FARSETTA et al. 1996; METCALF et al. 1991), and has RNA polymerase activity (DRAYNA and FIELDS 1982; STARNES and JOKLIK 1993). The M1-encoded μ2 protein lies adjacent to λ3 (FARSETTA et al. 1996), and genetic analyses have identified the M1 gene as a determinant of RNA synthesis (COOMBS 1996; SHERRY et al. 1996; YIN et al. 1996). In addition, we have recently demonstrated that μ2 is an RNA-binding protein (BRENTANO et al., submitted). Thus, the M1, L1, and L2 genes encode proteins involved in RNA synthesis. Together with the genetic analyses described above the data suggest that viral RNA synthesis is a determinant of reovirus-induced myocarditis.

4 Primary Cardiac Myocyte Cultures Provide an In Vitro Correlate to Investigate Reovirus-Induced Myocarditis

Given that reovirus-induced myocarditis is not immune-mediated, and that viral core proteins involved in RNA synthesis are determinants of the disease, we asked

	GENE SEGMENT DERIVATION										
	OUTER CAPSID			CORE					NON-STRUC		% MICE WITH GROSS
VIRUS	S1	S4	M2	S2	M1	L1	L2	L3	S3	M3	MYOCARDITIS
8B	3	1	3	1	1	1	1	1	1	1	100
EB121	1	3	1	3	3	3	3	1	3	3	0
E3	3	3	1	3	3	3	3	3	3	3	0
EW29	3	1	1	3	3	3	3	1	1	1	0
EW43	3	1	1	1	3	3	3	1	3	1	0
EW27	1	1	3	1	3	3	1	1	3	1	0
DB93A	1	1	3	3	3	1	3	3	1	3	0
DB93B	1	1	3	3	3	1	3	3	3	1	0
DB188	1	3	3	1	3	1	3	1	3	1	0
EW50	1	3	3	3	3	1	1	1	3	3	0
DB95	3	3	3	3	3	1	3	3	1	3	0
EW116	1	1	1	1	3	1	3	1	3	3	0
EW46	1	1	3	3	3	1	3	1	3	3	0
EW90	3	1	3	1	3	1	3	1	1	1	0
DB68	1	1	3	3	3	1	1	1	1	1	6
EW10	1	1	3	1	1	3	3	1	3	1	0
EW26	1	1	3	3	1	3	3	1	1	1	0
EW102	3	1	1	1	1	3	3	1	1	1	0
EW38	3	3	1	1	1	3	3	1	1	3	0
DB76	1	3	3	1	1	1	3	3	1	3	0
EW47	3	1	3	1	1	1	3	1	1	3	38
EW40	1	3	3	3	1	1	3	1	1	3	41
EW100	3	1	3	1	1	3	3	1	3	3	49
DB69A	1	1	3	3	1	3	1	3	1	1	56
DB62	1	3	3	1	1	1	3	1	1	1	57
EW25	3	1	3	1	1	1	3	1	1	1	71
EW67	3	1	1	1	1	1	1	1	1	1	75
EW93	3	1	1	1	1	3	1	1	1	1	80
EW96	1	1	3	1	1	1	1	1	3	3	80
EW112	3	1	3	1	1	1	3	1	3	1	83
DB181	1	3	3	1	1	1	1	1	1	1	89
EW89	3	1	3	3	1	1	3	1	1	3	100
EW60	1	1	3	1	1	1	1	1	1	1	100
DB88	3	1	3	1	1	1	3	1	3	3	100

Fig. 2. Genetic analysis of 8B-induced myocarditis: frequency of gross myocarditis. Neonatal Cr:(NIH)S mice were injected intramuscularly with 4×10^6–5×10^7PFU virus. Hearts were examined when the animals died or at 14 days postinjection, and a heart was scored as positive for gross myocarditis if any external lesions were detected. Gene segment derivations are indicated in the following two ways: *black/white box*, derivation from the myocarditic/nonmyocarditic parent, respectively; *1/3*, T1L/T3D origin of the gene segment. The EW series of viruses were derived from a cross between viruses 8B and EB121, and thus the origin of the L3 segment in EW viruses could not be determined. The DB series of viruses were derived from a cross between viruses EW60 and E3 (allowing determination of the origin of their L3 segments). (Adapted from Sherry and Blum 1994; Sherry et al. 1989)

whether infected primary cardiac myocyte cultures provide an in vitro correlate for viral myocarditis. First, we asked whether myocarditic reoviruses are more cytopathogenic to primary cardiac myocyte cultures than nonmyocarditic reoviruses, and if so, whether this cytopathogenicity is cell type specific. We generated primary cardiac myocyte and cardiac fibroblast cultures from fetal and neonatal mice and infected them with myocarditic and nonmyocarditic reoviruses (Baty and Sherry 1993).

Cytopathic effect (CPE) in cardiac myocytes 3 days postinfection was correlated with viral myocarditic potential ($p = 0.003$), but, interestingly, CPE in cardiac fibroblasts was not ($p > 0.05$). We later found that all of the viruses infect only a fraction of the cells during primary infections, and that progeny virus are generated in less than 1 day and can spread through the culture (Sherry et al. 1996). Thus by 3 days postinfection CPE reflects several rounds of infection. Therefore we compared CPE from primary infections (20h postinfection) with CPE following virus spread (5 days postinfection) (Sherry et al., in press). CPE was evident in primary infections, with viruses killing between 4% and 33% of the cells, but it was not correlated with viral myocarditic potential ($p > 0.05$). Instead, this CPE was associated with the S1 gene ($p = 0.001$), perhaps reflecting the efficiency of infection of cardiac myocytes (see Sect. 5.2).

In contrast, after viral spread cell death ranged from 18% to 75%, and CPE was correlated well with myocarditic potential ($p = 0.004$). While this CPE was

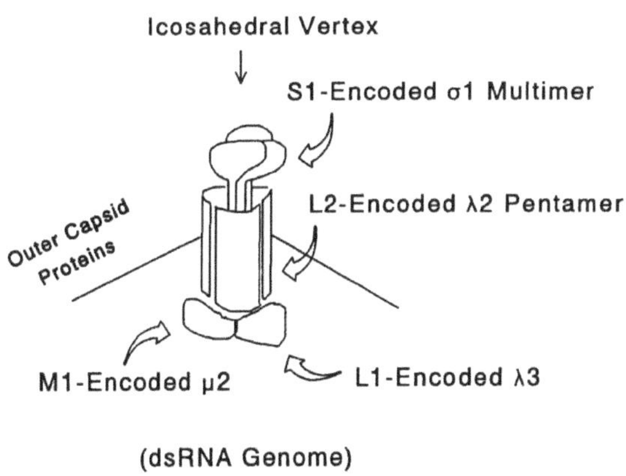

Fig. 3. Location of proteins in the reovirion associated with induction of gross myocarditis. (Based on data from Dryden et al. 1993; Farsetta et al. 1996; Metcalf et al. 1991; drawing adapted from Sherry and Blum 1994)

associated with the S1 gene ($p = 0.002$), it was also associated with the M1 gene ($p = 0.004$), which is a determinant of myocarditis. Together the data suggest that both myocarditic and nonmyocarditic reoviruses kill cardiac myocytes, but that myocarditic reoviruses spread more efficiently and thus induce greater cumulative cell death. The cardiac fibroblast results (BATY and SHERRY 1993) suggest that cumulative cell death, reflecting viral spread, is cell type specific.

5 Reovirus-Induced Myocarditis is Correlated with Viral RNA Synthesis Rather than Generation of Infectious Virus in Cardiac Myocytes

5.1 Reovirus-Induced Myocarditis is Correlated with Viral RNA Synthesis in Cardiac Myocytes

We next investigated viral replication in cardiac myocyte cultures to identify parameters that determine viral spread and cell death, and that are correlated with viral myocarditic potential. Since viral genes involved in RNA synthesis are correlated with viral myocarditic potential, we used Northern analyses to quantitate viral positive- and negative-strand RNA synthesis in cardiac myocyte cultures 10h postinfection using a large panel of myocarditic and nonmyocarditic reassortant reoviruses (SHERRY et al. 1996). Since the viruses could (and did; see below, and SHERRY et al. 1996) vary in their infection efficiencies, and the number of infected cells affects RNA measurements, we also calculated the ratio of positive- to negative-strand synthesis to measure RNA synthesis independent of infection efficiency. Both positive-strand RNA synthesis and the ratio of positive- to negative-strand RNA synthesis are correlated with viral myocarditic potential ($p = 0.036$ and 0.010, respectively; SHERRY et al. 1996). The S1 and M1 genes were associated with RNA synthesis, consistent with the S1 gene determining the efficiency of infection of cardiac myocytes (Sect. 5.2) and with COOMBS and colleagues evidence for M1 gene involvement in RNA synthesis (COOMBS 1996; YIN et al. 1996).

Together these data indicate that myocarditic viruses synthesize more viral RNA than nonmyocarditic viruses do by 10 h postinfection. We repeated the experiments in primary cardiac fibroblast cultures and mouse L929 cells (unpublished data), and viral RNA synthesis in both was correlated with viral myocarditic potential. Importantly, while CPE in cardiac myocytes was correlated with viral myocarditic potential (Sect. 4; BATY and SHERRY 1993), CPE in cardiac fibroblasts (BATY and SHERRY 1993) and mouse L929 cells (unpublished data) did not. Thus the data suggested that cell-specific factors do not determine RNA synthesis, but instead determine the consequences of RNA synthesis. The consequences in virally infected cardiac myocyte cultures are CPE, and this is correlated with viral myocarditic potential.

5.2 Efficiency of Infection and Yield of Infectious Virus from Primary Infections of Cardiac Myocytes do not Determine Myocarditis

To determine the efficiency with which reoviruses infect cardiac myocytes we used immunocytochemistry to identify infected cells in myocyte cultures infected with our large panel of viruses (SHERRY et al. 1996). While the efficiency of infection varied between viruses, it was correlated with the S1 gene ($p = 0.004$), encoding the viral attachment protein rather than viral myocarditic potential ($p > 0.05$). Since viral RNA synthesis is correlated with viral myocarditic potential, we asked whether viral yield is as well.

We infected cardiac myocyte cultures with our large panel of viruses and harvested duplicate cultures for immunocytochemistry and quantification of plaque-forming units (PFU) from primary infections (adding a viral inhibitor after primary infections, precluding spread of virus; SHERRY et al. 1996). The results were combined to determine the PFU per infected cell for each virus. Most of the viruses generated similar yields of infectious virus per infected cell, and thus yield was not correlated with induction of myocarditis or any reovirus gene ($p > 0.05$). Thus, while RNA synthesis in cardiac myocytes was correlated with myocarditic potential, viral RNA synthesis did not determine viral yield from infected cells.

6 Spread Between Cardiac Myocytes Correlates with Viral Myocarditic Potential, and Interferon-α/β is a Determinant of that Spread

As described above, viral RNA synthesis during initial (primary) infections of primary cardiac myocyte cultures and viral spread and cumulative cell death was correlated with viral myocarditic potential, irrespective of viral yield from the primary infections and irrespective of cell death during the primary infections. One mechanism by which viral RNA synthesis can determine viral spread is by induction of interferon (IFN; reviewed in VILCEK and SEN 1996). IFN is induced by viral dsRNA and is then secreted from infected cells. Upon binding neighboring cells IFN stimulates the synthesis of multiple proteins which directly exert the antiviral effect. Interestingly, these antiviral proteins are latent until activated (directly or indirectly) by dsRNA, suggesting that various patterns of viral RNA synthesis affect both induction of and sensitivity to IFN.

To test whether IFN is a determinant of reovirus spread between cardiac myocytes we used anti-IFN-α/β antibody to neutralize IFN in the cultures and then quantitated the effect on viral yield (SHERRY et al., in press). Specifically, we infected the cultures at 0.1PFU per cell, added control antibody or anti-IFN-α/β antibody, incubated for 7 days to allow viral spread, and then quantitated virus yield by plaque assay. Viral yield from control cultures varied three logs between viruses and was correlated with viral myocarditic potential ($p < 0.001$), indicating again that spread between myocytes is a determinant of myocarditis. This viral yield was associated with the M1 and L2 genes ($p < 0.001$ and $p = 0.049$, respectively), consistent with our earlier studies using T1L/T3D reassortant viruses (MATOBA et al. 1991). Anti-IFN-α/β antibody enhanced every virus' replication; however, the benefit from anti-IFN-α/β antibody varied three logs between viruses and was correlated with viral myocarditic potential (Fig. 4) ($p < 0.001$).

To determine whether induced IFN-α/β inhibits reovirus spread in other cell types we repeated the experiment in a differentiated skeletal muscle cell line (C2C12 cells) (BLAU et al. 1983). In these differentiated muscle cells viral yield in control cultures varied only 1.5 logs between viruses and was not correlated with viral myocarditic potential (SHERRY et al., in press). Moreover, the benefit from anti-IFN-α/β antiserum was maximally threefold, and the benefit was not correlated with viral myocarditic potential (Fig. 4). Therefore the role of IFN-α/β in controlling reovirus spread was cell type-specific, and in cardiac myocytes nonmyocarditic viruses induced more IFN-α/β and/or were more sensitive to the antiviral effects of IFN-α/β than myocarditic viruses.

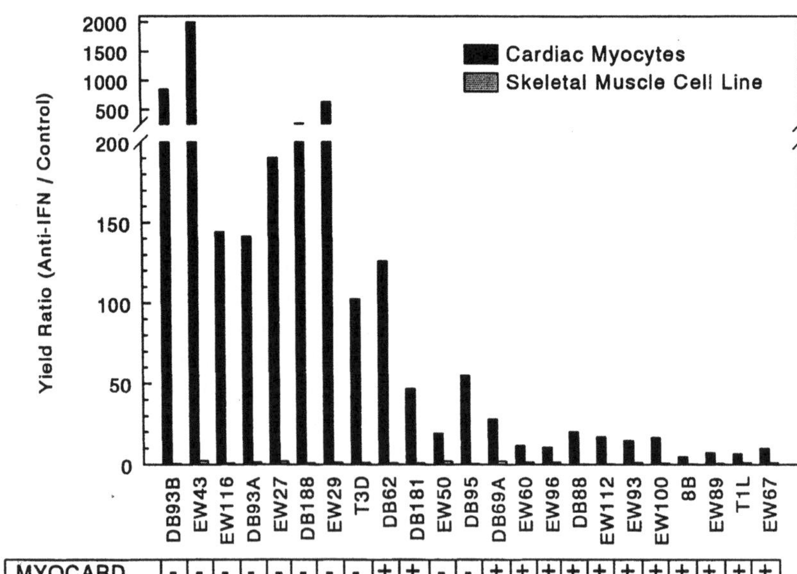

Fig. 4. Effect of anti-IFN-α/β antibody on reovirus spread through cardiac myocyte and differentiated skeletal muscle cell cultures. Primary cardiac myocyte cultures prepared from fetal Cr:(NIH)S mice (BATY and SHERRY 1993) and differentiated C2C12 cells (a murine skeletal muscle cell line, BLAU et al. 1983) were infected at a MOI of 0.1 PFU per cell, and anti-IFN-α/β antibody or control rabbit antiserum was added. Seven days after infection, yield of infectious virus from cultures receiving anti-IFN-α/β antibody was expressed as a ratio relative to cultures receiving control antibody, providing a measure of the benefit of the anti-IFN-α/β antibody to virus replication. (Adapted from SHERRY et al., in press)

7 Both Induction of Interferon-β and Sensitivity to the Antiviral Effects of Interferon-α/β in Cardiac Myocytes are Determinants of Reovirus Myocarditic Potential

To determine whether the anti-IFN-α/β antibody benefit to viral spread reflects induction of IFN-α/β, cardiac myocyte cultures were infected with our panel of viruses, and culture supernatants were titered at 10 and 20h postinfection for IFN-α/β levels by bioassay (SHERRY et al., in press). At 10h postinfection 8 of 12 nonmyocarditic viruses had induced detectable IFN while only 2 of 15 myocarditic viruses had done so. The magnitude of IFN induction was correlated with viral myocarditic potential ($p = 0.006$) and was associated with the M1 ($p = 0.040$), S2 ($p = 0.004$), and L2 ($p = 0.037$) genes. At 20h postinfection nearly all (22/27) viruses induced detectable IFN, but again the magnitude of IFN induction was correlated with viral myocarditic potential ($p = 0.007$). This IFN induction was associated again with the M1 ($p = 0.038$), S2 ($p = 0.012$), and L2 ($p = 0.044$) genes. Thus nonmyocarditic viruses induced more IFN than myocarditic viruses did in cardiac myocytes, and IFN induction was associated predominantly with genes encoding viral core proteins.

To distinguish between IFN-α and IFN-β, RNA was harvested from infected cardiac myocyte cultures for RT-PCR, using primers specific for IFN-α, IFN-β, or the control G3PDH (SHERRY et al., in press). IFN-β mRNA was amplified from nonmyocarditic viral infections but was not detectable from myocarditic viral or mock infections. In contrast, IFN-α mRNA was not amplified from any culture, despite the strong positive signal from a DNA control. Thus reoviruses induced IFN-β in cardiac myocytes.

To determine whether the benefit of anti-IFN-α/β antibody to viral spread also reflects sensitivity to IFN-α/β, cardiac myocyte cultures were pretreated with IFN-α/β or control buffer, then challenged with a panel of viruses, and harvested at 20h postinfection to determine viral yield (SHERRY et al., in press). IFN-α/β treatment inhibited each virus (maximally 2 logs), indicating that all viruses were sensitive to IFN-α/β. Despite the limited variation between viruses (1 log) IFN-α/β inhibition was correlated with viral myocarditic potential ($p = 0.043$) and was associated with the M1 gene ($p = 0.043$). To maximize sensitivity we modified the experiment by infecting at a low MOI and incubating the cultures for 7 days instead of 20h. In addition, since the viruses vary in their IFN-α/β induction, we included anti-IFN-α/β antibody in the control cultures. IFN-α/β inhibited viral yield 2–6.5 logs, and this inhibition was correlated with myocarditic potential ($p = 0.003$), and was associated with the M1 and L2 genes ($p = 0.016$ and 0.033, respectively). Thus nonmyocarditic reoviruses were more sensitive to the antiviral effects of IFN-α/β than myocarditic viruses, and this sensitivity was determined by genes encoding viral core proteins.

To determine directly whether IFN-α/β is a determinant of myocarditis we depleted IFN-α/β in Cr:NIH(S) neonatal mice using anti-IFN-α/β antibody (or control antibody) and then injected them with a nonmyocarditic virus that both induces IFN-α/β and is sensitive to IFN-α/β (DB188, SHERRY et al., in press). Mice were sacrificed 7 days postinjection and cardiac sections were examined for lesions. There were no lesions in the cardiac sections from control antibody-treated DB188-infected mice but multiple lesions in cardiac sections from the anti-IFN-α/β antibody treated DB188-infected mice. Thus a nonmyocarditic virus induced cardiac lesions when IFN-α/β was depleted, demonstrating that IFN-α/β is a determinant of its myocarditic potential. Since the antibody neutralized an unknown fraction of the IFN-α/β, and it is uncertain whether neutralization was uniform throughout the tissues (including the heart), it is likely that complete and uniform depletion, using IFN-α/β-receptor knockout mice (MULLER et al. 1994) would magnify the effects seen here, and those experiments are in progress.

8 Conclusions

Reovirus provides an ideal tool for investigating the mechanisms of nonimmune-mediated myocarditis. Results are summarized in Fig. 5, as follows. The efficiency with which reoviruses infect cardiac myocytes is determined by the S1 gene, most likely reflecting the function of the viral attachment protein that it encodes. Synthesis of viral RNA is determined by both the S1 and M1 genes and is correlated with induction of myocarditis. Generation of infectious virus, however, is corre-

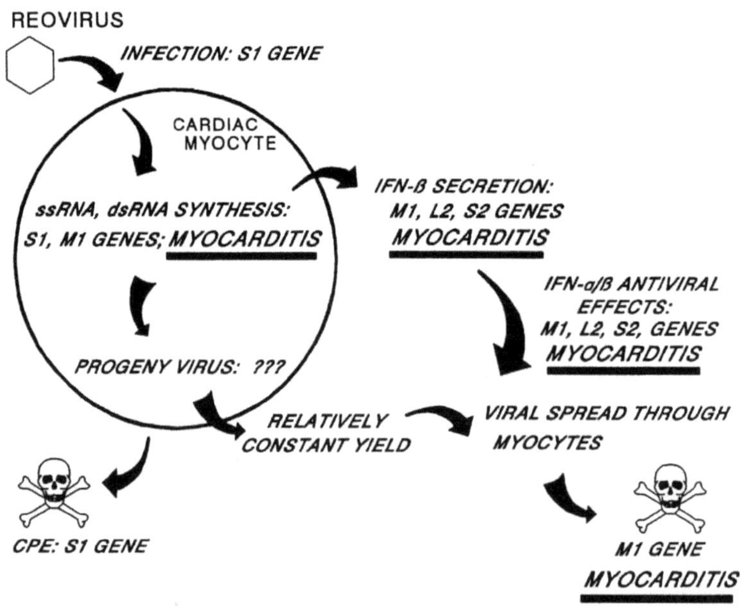

Fig. 5. Determinants of reovirus-induced acute myocarditis. (See Sect. 8)

lated with the fraction of cells infected and, when normalized to account for differences in infection efficiency, is relatively constant between viruses. Thus viral yield from initial (primary infections) is not correlated with viral RNA synthesis, induction of myocarditis, or any reovirus gene.

All of the reoviruses can kill cardiac myocytes; however, cytopathic effect in the initial infections is correlated with the S1 gene and the fraction of cells that are infected rather than induction of myocarditis. In contrast, viral spread between cardiac myocytes is well correlated with induction of myocarditis, and both cumulative cardiac myocyte cell death and cumulative viral yield from multiple rounds of infection are correlated with viral myocarditic potential. Critically, both induction of IFN-β and sensitivity to IFN-α/β are determinants of this spread, and nonmyocarditic viruses both induce more IFN-β and are more sensitive to IFN-β than myocarditic viruses. This provides the first direct evidence that IFN-β induction and IFN-α/β sensitivity can determine viral myocarditis.

Interestingly, IFN-α treatment of a patient with enterovirus-induced myocarditis which progressed to dilated cardiomyopathy was beneficial (Heim et al. 1994). Moreover, IFN-α treatment has recently been demonstrated to be beneficial for patients with idiopathic myocarditis or dilated cardiomyopathy (Miric et al. 1996), suggesting that the IFNs may play an important role in other viral myocardities as well. Continued investigations of reovirus-induced myocarditis are likely to provide additional insights into the role of IFN and the mechanisms by which viruses induce this important human disease.

Acknowledgements. B.S. is indebted to the members of her laboratory and to her reovirus colleagues for their continued support.

References

Archard LC, Bowles NE, Cunningham L, Freeke CA, Olsen EGJ, Rose ML, Meany B, Why HJF, Richardson PJ (1991) Molecular probes for detection of persisting enterovirus infection of human heart and their prognostic value. Eur Heart J 12:56–59

Aretz HT, Billingham ME, Edwards WD, Factor SM, Fallon JT, Fenoglio JJ, Olsen EGJ, Schoen FJ (1986) Myocarditis: a histopathologic definition and classification. Am J Cardiovasc Pathol 1:3–14

Bandt CM, Staley NA, Noren GR (1979) Acute viral myocarditis: clinical and histological changes. Minn Med 62:234–237

Baty CJ, Sherry B (1993) Cytopathogenic effect in cardiac myocytes but not in cardiac fibroblasts is correlated with reovirus-induced acute myocarditis. J Virol 67:6295–6298

Blau HM, Choy-Pik C, Webster C (1983) Cytoplasmic activation of human nuclear genes in stable heterokaryons. Cell 32:1171–1180

Borggrefe M, Block M, Breithardt G (1994) Identification and management of the high risk patient with dilated cardiomyopathy. Br Heart J [Suppl] 72:S42–S45

Bowles NE, Richardson PJ, Olsen EGJ, Archard LC (1986) Detection of Coxsackie-B-virus-specific RNA sequences in myocardial biopsy samples from patients with myocarditis and dilated cardiomyopathy. Lancet 1(8490):1120–1122

Cherry JD (1995) Enteroviruses. In: Remington JS, Klein JO (eds) Infectious diseases of the fetus and newborn infant. Saunders, Philadelphia, pp 404–446

Chow LH, Beisel KW, McManus BM (1992) Enteroviral infection of mice with severe combined immunodeficiency. Evidence for direct viral pathogenesis of myocardial injury. Lab Invest 66:24–31

Cleveland DR, Zarbl H, Millward S (1986) Reovirus guanylyltransferase is L2 gene product lambda 2 protein. J Virol 60:307–311
Coen D, Ramig RF (1996) Viral genetics. In: Fields BN, Knipe DM, Howley PM (eds) Fields virology. Lippincott-Raven, Philadelphia, pp 113–152
Cook DN (1996) The role of MIP-1 alpha in inflammation and hematopoiesis. J Leukoc Biol 59:61–66
Cook DN, Beck MA, Coffman TM, Kirby SL, Sheridan JF, Pragnell IB, Smithies O (1995) Requirement of MIP-1 alpha for an inflammatory response to viral infection. Science 269:1583–1585
Coombs K (1996) Identification and characterization of a double-stranded RNA reovirus temperature-sensitive mutant defective in minor core protein mu2. J Virol 70:4237–4245
Di Lenarda A, Secoli G, Perkan A, Gregori D, Lardiere G, Pinamonti B, Sinagra G, Zecchin M, Camerini F (1994) Changing mortality in dilated cardiomyopathy. Br Heart J 72:S46–S51
Drayna D, Fields BN (1982) Activation and characterization of the reovirus transcriptase: genetic analysis. J Virol 41:110–118
Dryden JA, Wang G, Yeager M, Nibert ML, Coombs KM, Furlong DB, Fields BN, Baker TS (1993) Early steps in reovirus infection are associated with dramatic changes in supramolecular structure and protein conformation: analysis of virions and subviral particles by cryoelectron microscopy and image reconstruction. J Cell Biol 122:1023–1041
Easton AJ, Eglin RP (1988) The detection of coxsackievirus RNA in cardiac tissue by in situ hybridization. J Gen Virol 69:285–291
Farsetta DL, Dryden KA, Nibert ML (1996) Identification of three proteins in reovirus top component particles that contribute to an internal structure seen by cryo-electron microscopy and image reconstruction. In: American Society for Virology annual meeting, London, Ontario
Goller T, Galle J, Eggers HJ, Bultmann B (1986) Experimental reovirus myocaritis in newborn mice: electron microscopic observations. Virchows Arch B Cell Pathol 50:373–386
Grun JB, Schultz M, Finkelstein SD, Crowell RL, Landau BJ (1988) Pathogenesis of acute myocardial necrosis in inbred mice infected with coxsackievirus B3. Microb Pathog 4:417–430
Guthrie M, Lodge PA, Huber SA (1984) Cardiac injury in myocarditis induced by coxsackievirus group B, type 3 in BALB/c mice is mediated by Lyt 2+ cytolytic lymphocytes. Cell Immunol 88:558–567
Haller BL, Barkon ML, Vogler GP, Virgin HW (1995) Genetic mapping of reovirus virulence and organ tropism in severe combined immunodeficient mice: organ-specific virulence genes. J Virol 69:357–364
Hashimoto I, Tatsumi M, Nakagawa M (1983) The role of T lymphocytes in the pathogenesis of coxsackievirus B3 heart disease. Br J Exp Pathol 64:497–504
Hassan SA, Rabin ER, Melnick JL (1965) Reovirus myocarditis in mice: an electron microscopic, immunofluorescent, and virus assay study. Exp Mol Pathol 4:66–80
Heim A, Stille-Siegner M, Kandolf R, Kreuzer H, Figulla HR (1994) Enterovirus-induced myocarditis: hemodynamic deterioration with immunosuppressive therapy and successful application of interferon-alpha. Clin Cardiol 17:563–565
Herzum M, Ruppert V, Kuytz B, Jomaa H, Nakamura I, Maisch B (1994) Coxsackievirus B3 infection leads to cell death of cardiac myocytes. J Mol Cell Cardiol 26:907–913
Hiraoka Y, Kishimoto C, Takada H, Nakamura M, Kurokawa M, Ochiai H, Shiraki K (1996) Nitric oxide and murine coxsackievirus B3 myocarditis: aggravation of myocarditis by inhibition of nitric oxide synthase. J Am Coll Cardiol 28:1610–1615
Huber SA (1992) Viral myocarditis – a tale of two diseases. Lab Invest 66:1–3
Huber SA, Heintz N, Tracy R (1988) Coxsackievirus B 3-induced myocarditis. Virus and actinomycin D treatment of myocytes induces novel antigens recognized by cytolytic T lymphocytes. J Immunol 141:3214–3219
Hufnagel G, Chapman N, Tracy S (1995) A non-cardiovirulent strain of coxsackievirus B3 causes myocarditis in mice with severe combined immunodeficiency syndrome. Eur Heart J 16:18–19
Kandolph R, Ameis D, Kirschner P, Canu A, Hofschneider PH (1987) In situ detection of enteroviral genomes in myocardial cells by nucleic acid hybridization: An approach to the diagnosis of viral heart disease. Proc Natl Acad Sci USA 84:6272–6276
Kandolph R, Klingel K, Mertsching H, Canu A, Hohenadl C, Zell R, Reimann Y, Heim A, McManus BM, Foulis AK, Schultheiss H-P, Erdmann E, Riecker G (1991) Molecular studies on enteroviral heart disease: patterns of acute and persistent infections. Eur Heart J 12:49–55
Kaplan MH, Klein SW, McPhee J, Harper RG (1983) Group B coxsackievirus infections in infants younger than three months of age: a serious childhood illness. Rev Infect Dis 5:1019–1032
Keeling PJ, Lukaszyk A, Poloniecki J, Caforio AL, Davies MJ, Booth JC, McKenna WJ (1994) A prospective case-control study of antibodies to coxsackie B virus in idiopathic dilated cardiomyopathy. J Am Coll Cardiol 23:593–598

Kishimoto C, Abelmann WH (1990) In vivo significance of T cells in the development of Coxsackievirus B3 myocarditis in mice Immature but antigen-specific T cells aggravate cardiac injury. Circ Res 67:589–598

Klingel K, Stephan S, Sauter M, Zell R, McManus BM, Bultmann B, Kandolf R (1996) Pathogenesis of murine enterovirus myocarditis: virus dissemination and immune cell targets. J Virol 70:8888–8895

Lerner AM, Cherry JD, Klein JO, Finland M (1947) Infections with reoviruses. N Engl J Med 267:947–952

Leslie K, Blay R, Haisch C, Lodge A, Weller A, Huber SA (1989) Clinical and experimental aspects of viral myocarditis. Clin Microbiol Rev 2:191–203

Liu P, Penninger J, Aitken K, Sole M, Mak T (1995) The role of transgenic knockout models in defining the pathogenesis of viral heart disease. Eur Heart J 16:25–27

Lowenstein CJ, Hill SL, Lafond-Walker A, Wu J, Allen G, Landavere M, Rose NR, Herskowitz A (1996) Nitric oxide inhibits viral replication in murine myocarditis. J Clin Invest 97:1837–1843

Mao ZX, Joklik WK (1991) Isolation and enzymatic characterization of protein lambda 2, the reovirus guanylyltransferase. Virology 185:377–386

Marboe CC, Fenoglio JJ (1988) Pathology and natural history of human myocarditis. Pathol Immunopathol Res 7:226

Martin A, Webber S, Fricker F, et al (1994) Acute myocarditis: a rapid diagnosis by PCR in children. Circulation 90:330–339

Martino T, Aitken K, Penninger J, Mak T, Sole MJ (1995) Mice with p56(ick) T-lymphocyte tyrosine kinase knocked out are resistant to coxsackieviral myocarditis (abstract from the 1995 American Heart Association meeting). Circulation 92 [Suppl I]:I-787

Martino TA, Liu P, Sole MJ (1994) Viral infection and pathogenesis of dilated cardiomyopathy. Circ Res 74:182–188

Matoba Y, Sherry B, Fields BN, Smith TW (1991) Identification of the viral genes responsible for growth of strains of reovirus in cultured mouse heart cells. J Clin Invest 87:1628–1633

Matsumori A, Kawai C (1982) An animal model of congestive (dilated) cardiomyopathy: dilatation and hypertrophy of the heart in the chronic stage in DBA/2 mice with myocarditis caused by encephalomyocarditis virus. Circulation 66:355–360

Matsumori A, Matoba Y, Sasayama S (1995) Dilated cardiomyopathy associated with hepatitis C virus infection. Circulation 92:2519–2525

Metcalf P, Cyrklaff M, Adrian M (1991) The three-dimensional structure of reovirus obtained by cryo-electron microscopy. EMBO J 10:3129–3136

Mikami S, Kawashima S, Kanazawa K, Hirata K, Katayama Y, Hotta H, Hayashi Y, Ito H, Yokoyama M (1996) Expression of nitric oxide synthase in a murine model of viral myocarditis induced by coxsackievirus B3. Biochem Biophys Res Commun 220:983–989

Miric M, Vasiljevic J, Bojic M, Popovic Z, Keserovic N, Pesic M (1996) Long-term follow up of patients with dilated heart muscle disease treated with human leucocytic interferon alpha or thymic hormones. Initial results. Heart 75:596–601

Morimoto S-I, Yamada K, Kubo N, Mizuno Y, Hasumi M, Hiramitsu S, Uemura A, Kimura K, Nishikawa T, Sekiguchi M (1992) Clinical and pathological features of chronic myocarditis: four autopsy cases presenting as dilated cardiomyopathy in life. Am J Cardiovasc Pathol 4:181–191

Muller U, Steinhoff U, Reis LFL, Hemmi S, Pavlovic J, Zinkernagel RM, Aguet M (1994) Functional role of type I and type II interferons in antiviral defense. Science 264:1918–1921

Nibert ML, Schiff LA, Fields BN (1996) Reoviruses and their replication. In: Fields BN, Knipe DM, Howley PM (eds) Fields virology. Lippincott-Raven, Philadelphia, pp 1557–1596

Oberhaus SM, Smith RL, Clayton GH, Dermody TS, Tyler KL (1997) Reovirus infection and tissue injury in the mouse central nervous system are associated with apoptosis. J Virol 71:2100–2106

Okuni M, Yamada T, Mochizuki S, Sakurai I (1975) Studies on myocarditis in childhood with special reference to the possible role of immunological process and the thymus in the chronicity of the disease. Jpn Circ J 39:463–470

Olsen EGJ (1992) The pathogenesis of dilated cardiomyopathy. Postgrad Med J 68:S7–S10

Rabin ER, Hassan SA, Jenson AB, Melnick JL (1964) Coxsackievirus B3 myocarditis in mice. An electron microscopic, immunofluorescent and virus-assay study. Am J Pathol 44:775–797

Rodgers SE, Barton ES, Oberhaus SM, Pike B, Gibson CA, Tyler KL, Dermody TS (1997) Reovirus-induced apoptosis of MDCK cells is not linked to viral yield and is blocked by Bcl-2. J Virol 71:2540–2546

Rose NR, Herskowitz A, Neumann DA, Neu N (1988) Autoimmune myocarditis: a paradigm of post-infection autoimmune disease. Immunol Today 9:118–120

Rose NR, Neumann DA, Herskowitz A (1992) Coxsackievirus myocarditis. Adv Intern Med 37:411–429
Schwimmbeck PL, Badorff C, Rohn G, Schultze K, Schultheiss HP (1996) The role of sensitized T cells in myocarditis and dilated cardiomyopathy. Int J Cardiol 54:117–125
See DM, Tilles JG (1991) Viral myocarditis. Rev Infect Dis 13:951–956
Sherry B, Blum MA (1994) Multiple viral core proteins are determinants of reovirus-induced acute myocarditis. J Virol 68:8461–8465
Sherry B, Fields BN (1989) The reovirus M1 gene, encoding a viral core protein, is associated with the myocarditic phenotype of a reovirus variant. J Virol 63:4850–4856
Sherry B, Torres J, Blum MA (1998) Reovirus induction of and sensitivity to beta interferon in cardiac myocyte cultures correlate with induction of myocarditis and are determined by viral core proteins. J. Virol In Press
Sherry B, Schoen FJ, Wenske E, Fields BN (1989) Derivation and characterization of an efficiently myocarditic reovirus variant. J Virol 63:4840–4849
Sherry B, Li X, Tyler KL, Cullen JM, Virgin HW (1993) Lymphocytes protect against and are not required for reovirus-induced myocarditis. J Virol 67:6119–6124
Sherry B, Baty CJ, Blum MA (1996) Reovirus-induced acute myocarditis in mice correlates with viral RNA synthesis rather than generation of infectious virus in cardiac myocytes. J Virol 70:6709–6715
Sole M, Liu P (1994) Viral myocarditis: a paradigm for understanding the pathogenesis and treatment of dilated cardiomyopathy. J Am Coll Cardiol 22:99A–105A
Stangl E, Aschauer W, Zahringer J, Hubner G (1987) Reovirus myocarditis. Eur Heart J 8 [Suppl J]:407–409
Starnes MC, Joklik WK (1993) Reovirus protein lambda 3 is a poly(C)-dependent poly(G) polymerase. Virology 193:356–366
Tracy S, Chapman NM, McManus BM, Pallansch MA, Beck MA, Carstens J (1990a) A molecular and serologic evaluation of enteroviral involvement in human myocarditis. J Mol Cell Cardiol 22:403–414
Tracy S, Wiegand V, McManus B, Gauntt C, Pallansch M, Beck M, Chapman N (1990b) Molecular approaches to enteroviral diagnosis in idiopathic cardiomyopathy and myocarditis. J Am Coll Cardiol 15:1688–1694
Tyler KL, Fields BN (1996) Reoviruses. In: Fields BN, Knipe DM, Howley PM (eds) Fields virology. Lippincott-Raven, Philadelphia, pp 1597–1624
Tyler KL, Squier MK, Rodgers SE, Schneider BE, Oberhaus SM, Grdina TA, Cohen JJ, Dermody TS (1995) Differences in the capacity of reovirus strains to induce apoptosis are determined by the viral attachment protein sigma 1. J Virol 69:6972–6979
Tyler KL, Squier MK, Brown AL, Pike B, Willis D, Oberhaus SM, Dermody TS, Cohen JJ (1996) Linkage between reovirus-induced apoptosis and inhibition of cellular DNA synthesis: role of the S1 and M2 genes. J Virol 70:7984–7991
Vilcek J, Sen GC (1996) Interferons and other cytokines. In: Fields BN, Knipe DM, Howley PM (eds) Fields virology. Lippincott-Raven, Philadelphia, pp 375–400
Virgin HW, Tyler KL, Dermody TS (1997) Reovirus. In: Nathanson N (ed) Viral pathogenesis. Lippincott-Raven, New York, pp 669–702
Walters MN-I, Leak PJ, Joske RA, Stanley NF, Perret DH (1965) Murine infection with reovirus. III. Pathology of infection with types 1 and 2. Br J Exp Pathol 46:200–212
Wang J, Walsh K (1996) Resistance to apoptosis conferred by Cdk inhibitors during myocyte differentiation. Science 273:359–361
Weiss LM, Movahed LA, Billingham ME, Cleary ML (1991) Detection of coxsackievirus B3 RNA in myocardial tissues by the polymerase chain reaction. Am J Pathol 138:497–503
Wenger NK, Abelmann WH, Roberts WC (1990) Myocarditis. In: Hurst JW, Schlant RC (eds) The heart, arteries, and veins. McGraw-Hill, New York, pp 1256–1277
Wenske EA, Chanock SJ, Krata L, Fields BN (1985) Genetic reassortment of mammalian reoviruses in mice. J Virol 56:613–616
Wolfgram LJ, Beisel KW, Rose NR (1985) Heart specific autoantibodies following murine coxsackievirus B3 myocarditis. J Exp Med 161:1112–1121
Woodruff JF (1980) Viral myocarditis, a review. Am J Pathol 101:427–479
Yin P, Cheang M, Coombs KM (1996) The M1 gene is associated with differences in the temperature optimum of the transcriptase activity in reovirus core particles. J Virol 70:1223–1227

Pathogenesis of Reovirus Gastrointestinal and Hepatobiliary Disease

E.L. ORGAN[1,2] and D.H. RUBIN[3,4]

1 Introduction

During the initial surveys in search of viral infectious agents carried out in the late 1940s to the 1960s reovirus was discovered to be a common infectious agent that was recovered from the feces of virtually all mammals (ROSEN and ABINANTI 1960; HRDY et al. 1979). In humans, exposure to the mammalian reovirus occurs in childhood, with a high proportion of the population having serological evidence of infection by young adulthood (LERNER et al. 1947; JACKSON et al. 1961; LEERS and

[1]Research Medicine, Veterans Affairs Medical Center, Nashville, TN 37212, USA
[2]Department of Microbiology and Immunology, Vanderbilt University School of Medicine, Nashville, TN 37240, USA
[3]Research Medicine, Veterans Affairs Medical Center, Nashville, TN 37212, USA
[4]Department of Medicine, Division of Infectious Diseases, and Department of Microbiology and Immunology, Vanderbilt University School of Medicine, Nashville, TN 37240, USA

ROYCE 1966). However, a role of reovirus in the pathogenesis of symptomatic infection has not been substantiated in humans. Although reoviruses have not been conclusively implicated in any human disease, this viral system has served as a model to study the manner by which other more pathogenic viruses in humans may infect the gut and gain entry into a mammalian host. The ability to manipulate the viral genetic elements has enhanced our ability to define viral proteins associated with virus delivery, entry, and infection at the primary site of entry and has been an additional rationale to study reovirus infection.

2 Initiation of Infection: A Role for Proteolytic Enzymes

The intestine, a site of primary contact with reovirus, is capable of supporting virus replication. In mammalian hosts the intestinal epithelium is infected in various ways, including oral inoculation, introduction by nasal spray, or intravenous injection (TYLER and FIELDS 1996; RUBIN et al. 1986a). Moreover, replication of virus and release into the feces provides a route of dissemination of infectious virus into the environment. Therefore the replication of reovirus in the intestine serves as a vehicle for entry and expansion of infectious virions and a site for egress for eventual infection of new hosts.

Closely packed reovirions that form at lattice array within the cell are released when cells die. In this form, the genetic material is protected from the outside world by two distinct concentric coats of proteins, and the virus attachment protein, σ1, is retracted to the virion surface (NIBERT et al. 1991, 1996). This structure of the virion survives in sewage or river water and is ingested by a second host to begin the infectious cycle anew. The virion is extremely stable in the intestinal milieu and resists inactivation by acid, bile salts, and proteases (NIBERT et al. 1996). However, the ingested reovirion is modified by the digestive protease chymotrypsin in the lumen of the small intestine to the intermediate (or infectious) subviral particle (ISVP) (SPENDLOVE et al. 1970; JOKLIK 1972; RUBIN and FIELDS 1980; BASS et al. 1990). The ISVP loses the σ3 protein and partially cleaves the μ1c protein, which results in the extension of the σ1 protein (JOKLIK 1972; DRYDEN et al. 1993).

The biological significance of the transformation from virions to ISVPs and the resultant increase in infectivity has been studied by using virions and *in vitro* generated ISVPs to infect L cells *in vitro*. Reovirus type 1, strain Lang (1/L) showed a twofold increase in infectivity of L cells when converted from virion to ISVPs while reovirus type 3, Dearing (3/D), showed a 90% decrease in infectivity when handled in the same way. Curiously, the hemagglutination capacity of the type 1 reovirus was unchanged upon conversion to ISVP while that of type 3/D increased approximately eightfold (NIBERT et al. 1995; VIRGIN and TYLER 1991). In part, the changes in infectivity have been mapped to the partial cleavage of the μ1c of reovirus type 1, which results in the slight increase in infectivity after conversion to ISVPs (RUBIN and FIELDS 1980; BASS et al. 1990). For reovirus type 3/D a portion of the σ1 protein containing the cell receptor binding site is cleaved leaving the hemagglutinin domain in the fiber portion more exposed. Thus, binding of virus to cellular receptors is reduced and hemagglutination capacity is enhanced. Therefore, the loss of infectivity of reovirus type 3 in an intestinal milieu is related to the loss of the virus attachment proteins' binding site to its cellular receptor but not its capacity to bind to and infect red cell precursors (Nibert et al. 1995). The relative susceptibility of reovirus type 3/D to proteolytic inactivation in the intestine is not conserved by other reovirus type 3 isolates, suggesting that this mutation occurred with adaptation of this strain to laboratory conditions.

The importance of the conversion of the virion to ISVP can also be demonstrated *in vivo*. When neonatal mice were inoculated perorally with radiolabeled type 1/L virions, all virus which was recovered from the lumen of the intestine by 30 min. postinoculation was observed to have been converted to ISVPs. This process may be monitored by electrophoresis and autoradiography of the radiolabeled virus recovered from the lumen of the intestine since the outer coat proteins of the virion, σ3 and μ1c, are present in virions but not ISVPs (Bodkin et al. 1989). In addition, the conversion of virion to ISVP can be greatly inhibited in neonatal mice by intragastric pretreatment of the mice with the protease inhibitors aprotinin and chymostatin. When mice are given these protease inhibitors before peroral inoculation of virus, much lower titers of virus are produced from the infected intestines than in mice inoculated with identical titers but pretreated only with saline. This inhibition of virion to ISVP conversion may also be observed with autoradiography as above. A confirmatory experiment in which mice were pretreated with protease inhibitors and then infected with *in vitro* generated ISVPs revealed no inhibition of to virus growth as measured by titers taken from infected intestines up to 24h postinoculation (Bass et al. 1990). Also, electron micrography of actual virus binding to cells in the intestine confirmed directly that ISVPs bind much better than intact virions (Amerongen et al. 1994). Clearly, the conversion of virions to ISVPs is an important early step in the reovirus intestinal infection process.

One may hypothesize that it is beneficial for intact virions to pass through the distal bowel without binding to the host cells, as this would facilitate passage to the outside environment. As the ISVPs are less stable than whole virions, their presence in feces as the dominant viral form would not ensure transmission to a new host. Therefore, the relatively protease free distal intestine would be less likely to be directly infected by reovirus. However, the difference in binding and internalization between virions and ISVPs is not absolute (Bass et al. 1990; Amerongen et al. 1994). This may indicate that the intact virion has at least some capacity to bind to M cells and infect intestinal cells.

3 Site of Intestinal Absorption: M Cells and Polarized Epithelium

Electron micrography has consistently established that reovirus is highly specific in its binding to the apical or luminal side of one particular cell type, the M cell (Wolf et al. 1981, 1983; BASS et al. 1988; AMERONGEN et al. 1994). These are specialized epithelial cells that make up about 10% of the dome epithelial layer that overlies the intestinal lymph nodules, or Peyer's patches, found largely in the ileum. These cells have distinct microfolds on their luminal side and are thus known as M cells. Their function involves endocytosis of large molecules from the lumen of the intestine and transport of these molecules in vesicles to the intercellular space where the molecules then become accessible to the cells of the immune system. Reovirus, as with a number of other pathogens, such as poliovirus, HIV, some bacteria (*Mycobacteria*, *Salmonella*), fungi (*Candida*), and parasites (Cryptosporidia), utilize this transepithelial transport process to enter the host (SHINER 1985).

With electron microscopy, dense clusters of type 1 reovirus particles can be seen adhering to the luminal side of M cells within 30min after the intestines of suckling mice are inoculated with reovirus. In these same micrographs the directly adjacent absorptive cells are nearly devoid of adherent virus particles. Moreover, within 1h of infection endocytosed virus can be found within M cell vesicles and a few are even visible in the extracellular space beneath M cells, presumably having been transported already entirely through the M cell (WOLF et al. 1981, 1983; BASS et al. 1988). Additional studies have identified the binding and endocytosis of reovirus type 3/D to occur at the apical side of the absorptive cell in suckling mice (WOLF et al. 1983). Also, in the youngest suckling mice, (2- to 6-day-old) both crypt and villus cells were observed to attach and internalize virus in contrast with older mice in which binding is to the basal lateral side exclusively (WOLF et al. 1987).

Additional evidence that absorptive cells are not the major type of cell that reovirus attaches to comes from work with freshly isolated intestinal epithelial cells. When the binding of reovirus to primary cultures of adult mouse intestinal crypt and villus cells was tested *in vitro*, it was found that reovirus types 1 and 2 bind only to the nonluminally exposed basolateral side of the cells while reovirus type 3/D do not bind at all (RUBIN 1987; WEINER et al. 1988). Thus it seems well established that the M cells overlaying the Peyer's patches are the primary site of attachment and invasion of reovirus type 1 in the older suckling and adult mice. However, in *in vitro* cultures of Caco-2, a human adenocarcinoma cell line that maintains polarization, reovirus virions of both types 1 and 3 bind to and internalize from the apical (luminal) side of these cells. Moreover, rat intestinal apical membrane vesicles also bind both reovirus types 1 and 3 (AMBLER and MACKAY 1991). These results, which conflict with those obtained with freshly isolated tissues, may reflect a lack of polarization of the reovirus receptors under culture conditions or differences of viral tropism among the different animal species. Further study of this question requires the identification of the cellular receptor(s) for reovirus.

In addition to providing sites for virus entry into the host, M cells are also likely to serve as targets for virus replication (BASS et al. 1988). BASS and coworkers were able to demonstrate that virus factories were present in M cells at 4h following infection, suggesting active virus replication. Alternatively, the presence of "factories" so early after virus inoculation may demonstrate the incorporation of aggregates of virions and not replication, with the dramatic decrease in M cell numbers that occur 4h postinfection being due to a cytotoxic effect resulting from the high virus inoculum. The authors speculate that the rapid die-off of infected M cells may benefit the host, as their rapid depletion may limit the amount of pathogen that can enter by that route. In any case, the damage to the host is not permanent as the M cell population returns to normal by 13 days after infection (BASS et al. 1988). RUBIN and coworkers (1985) using immunohistochemical staining to demonstrate reovirus infected intestinal epithelial cells did not find evidence of M cell infection at 48h following peroral inoculation of reovirus type 1. However, murine M cells would be virtually impossible to visualize by light microscopy. Therefore the results obtained by these two laboratories are not necessarily conflicting.

4 Intestinal Epithelial Cell Infection

Using immunoperoxidase staining of infected adult mouse intestines, the crypt and dome absorptive epithelial cells were the initial reovirus antigen positive cells, at 48h, and mononuclear cells beneath the dome epithelium and in Peyer's patches were occasionally also antigen positive (RUBIN et al. 1985). There is considerable evidence that the infection of the absorptive cells is from the basolateral side of the cells. First, *in vitro*, isolated epithelial cells bind reovirus type 1 to their basolateral surface, the side accessible from the extracellular space (RUBIN 1987; WEINER et al. 1988). BASS and coworkers also found reovirus type 1 in the dome epithelium, villus absorptive cells, and some mononuclear cells 4–6h after infection in 10-day-old mice by electron micrography (BASS et al. 1988). They also found that the ratio of virus attached to the basal versus the apical side of the cell was 10:1. When this ratio was normalized to account for the greater surface area of the apical side of the absorptive cells the ratio was closer to 100:1 (BASS et al. 1988). Thus, the binding of virus is markedly greater to the basal rather than luminal side of absorptive cells. Third, these absorptive cells are found to become infected in animals that are injected intravenously with reovirus type 1, which exposes the basal side of the cells to virus (RUBIN et al. 1986a,b). Thus, the infection of the intestinal cell types found adjacent to the site of entry of the virus can be accounted for by viral tropism and anatomical proximity, supporting the contention that the reovirus type 1 receptor on murine intestinal epithelial cells is polarized.

The ileum is the site of the most intensive infection within the bowel. When adult mice were orally inoculated with a high titer (10^8 PFU, of reovirus type I) and the intestinal segments of duodenum, jejunum, ileum, and colon were titered for viral growth, virus was detected in all the segments. However, the titers of virus in the ileum were consistently equal to or greater than that in any of the other segments. In fact, by day 6 postinfection, the titer from the ileum was nearly three orders of magnitude greater than that from the other sections. Also, it was only in the ileum that the reovirus content in infected cells was sufficient for virus to be visualized by immunohistochemistry. Moreover, when adult mice were inoculated with progressively higher titers of virus and their bowels were examined microscopically for pathological changes, the evident damage was restricted to ileum.

At 2–14 days following infection mice that received 10^8–10^9 PFU reovirus had evidence of inflammatory cells in the lamina propria of the ileum. In addition, there was epithelial cell hyperplasia of the crypts with a loss of nuclear polarity. When the mice were given the extremely high dose of 10^{10} PFU per mouse, half of them (5/10) died by day 5. All of the dead mice had transmural perforations of the bowel restricted to the ileum (Rubin et al. 1985). Strikingly, when the mice were given the lethal dose of 10^{10} PFU of reovirus type 1 intravenously instead of orally, histopathological lesions in the intestine were limited to the ileum (Rubin et al. 1986a,b). These studies demonstrate that there is a high degree of tropism in reovirus type 1 for the ileum.

Within the ileum the cells most intensely infected were the undifferentiated cells in the crypts of Lieberkühn. While a few macrophages and some dome epithelium and villus cells stained for virus antigen, most of the reovirus antigen positive cells were in the crypts (Rubin et al. 1985). The difference between the infection of the crypt and villus cells is probably not due to their accessibility by the virus as the villus cells are supplied by greater vascularity than the crypts. However, the difference in viral growth may be due to the replicative capacity of the cells. Thymidine incorporation studies have shown that the zone of mitotic activity is limited to the crypts of Lieberkühn. By the time the villus cells have migrated from the crypts on their 2- to 6-day trip to the tips of the villi and eventual sloughing off, they have ceased to divide and have begun to differentiate. Thus reovirus, which grows best in rapidly dividing cells, may find the crypt cells to be better hosts (Duncan et al. 1978).

5 Intestinal Infection Occurs in Proliferating Cells

Direct experimental evidence that reovirus replication occurs in actively proliferating intestinal cells comes from *in vitro* work using rat intestinal epithelial cells (RIE-1 cells) (Blay and Brown 1984, 1985). When RIE-1 cells form a confluent monolayer, they cease to divide and partially differentiate into villuslike cells. Confluent monolayers of RIE-1 cells arrest cell development at the G_0-G_1 stage and reovirus type 1 infection is restricted to a few scattered cells as is evident by immunohistochemistry staining for virus antigen (Organ et al., submitted). However, if a quiescent monolayer of RIE-1 cells is scratched with a pipette tip, cells at the edge of the scratch proliferate until the gap in the monolayer is healed. If a scratched monolayer is infected with reovirus and then examined by immunohistochemistry for reovirus antigen as the gap is healed, the proliferating cells at the edge of the gap contain enough reovirus antigen to see the healed scratch against a relatively unstained background (Organ et al., submitted).

Together the above data support and enhance the model of reovirus type 1 intestinal infection in adult mice first proposed by Rubin et al. (1985) (Fig. 1). In this model the ingested virion is converted to an ISVP by the digestive enzymes of the host, and it then breaches the mucosal barrier of the intestine by passing through the M cell to the extracellular space. Then it either is endocytosed by a macrophage, which it infects, or it is taken up by a vein or the lymph system where it diffuses to the basal side of a dome epithelia or villus or crypt cell where the virion initiates an infection by being endocytosed. A productively infected crypt cell continues to migrate up towards the villus. Virus-laden villus cells are sloughed off at the villus tip or die prematurely, thereby releasing the virus into the lumen of the intestine. The released virus may then be converted into an ISVP and repeat the cycle. However, since it appears that most of the virus is produced in the ileum, which is distal to the pancreatic proteases, most of the virus is likely to retain its virion form and get passed from the body into the environment where it awaits ingestion to begin the cycle anew (Fig. 1).

Reovirus type 3 strains differ from reovirus type 1 in their capacity to replicate in the intestine. Both reovirus types 1 and 3 are endocytosed by M cells in suckling and

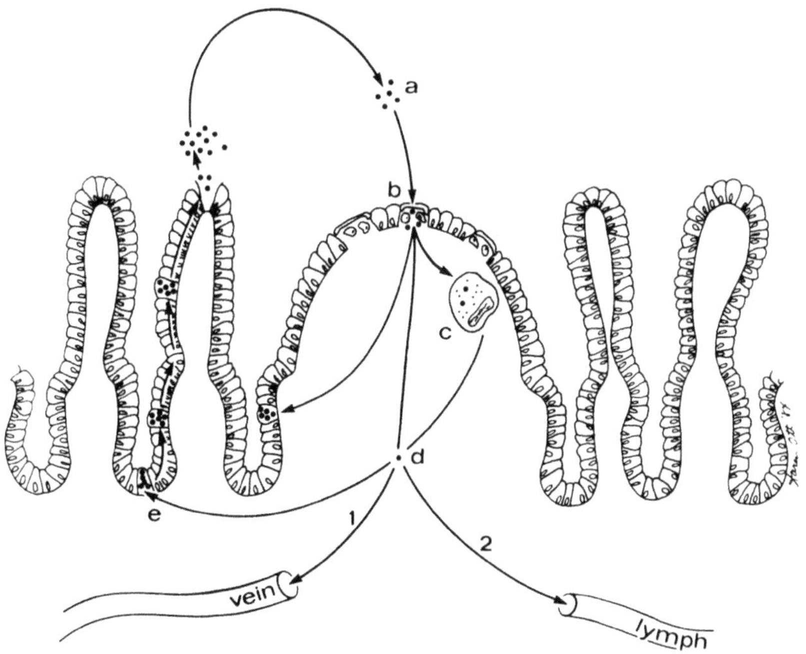

Fig. 1. Schematic diagram, showing the probable route of entry and local or systemic spread of reovirus. Virus (*a*) in the lumen of the intestine is transported into the host via M cells (*b*) to the macrophage-rich subepithelial region overlying a Peyer's patch. From the lymphoid follicle, the virus either is endocytosed (*c*) and carried intracellularly or follows freely in lymph to the basal crypt epithelial cells. Dissemination of virus may occur (*d*) by vein (*1*) or by lymphatic efferents (*2*) to the blood stream where it is circulated to systemic or intestinal sites supportive of virus replication, thereby resulting in infection of crypt cells distant to the Peyer's patch. In the intestine the infected basal crypt cell (*e*) migrates toward the villus tip during maturation into an absorptive enterocyte; lysis or exfoliation releases free virus into the intestinal lumen. (Reprinted from Rubin et al. 1985 with permission of the authors and the Journal of Virology)

adult mice. However, while reovirus type 3 is also apically endocytosed by absorptive cells in sucking mice, reovirus type 1 is not (WOLF et al. 1981, 1983; RUBIN et al. 1985; MORRISON et al. 1991; BASS et al. 1988). While some authors have described reovirus type 3 intestinal infection in suckling mice (KAUFFMAN et al. 1983; BRANSKI et al. 1980; KEROACK and FIELDS 1986; BODKIN and FIELDS 1989), the data are not entirely consistent. For example, some authors describe rapid clearance of reovirus from suckling mouse intestine (BODKIN et al. 1989) while others describe prolonged infection (BRANSKI et al. 1980). Differences in mouse strains and methodologies may explain some of the inconsistencies while differences in reovirus type 3 isolate clones may account for others (KEROACK and FIELDS 1986). It is also possible that prior to gut closure, the capacity of reoviruses to enter the apical side of intestinal epithelial cells exists (WALKER 1979; REINHARDT 1984). In adult mice there is very little evidence that reovirus type 3 infects intestinal cells at all unless enormous titers of virus are used to initiate infection (RUBIN et al. 1986b). One group (WEINER et al. 1988) using radiolabeled virus or indirect immunofluorescence has shown that reovirus type 3 fails to bind to isolated adult mouse intestinal epithelial cells. However, a tissue culture of intestinal cells derived from human cells, Caco-2 cells, is able to bind reovirus type 3 as are isolated rat intestinal cell vesicles (AMBLER and MCKAY 1991). Whether species differences result in the variation observed is uncertain.

6 Viral Determinants for Intestinal Infection and Systemic Spread

Reovirus isolates that have a reassortment of reovirus types 1 and type 3 gene segments have been used to map the tropism of type 1 to the intestine. Consistently, the S1 gene of type 1 which encodes the σ1 protein has been identified as an intestinal tropic factor (RUBIN et al. 1986b; BODKIN et al. 1989; HALLER et al. 1995a). Additionally, the M2 and L2 genes have been identified as playing a role in reovirus type 1 virulence in or transmission of intestinal infection (RUBIN and FIELDS 1980; KEROACK and FIELDS 1986; BODKIN et al. 1989). Both the S1 and L2 gene segment have been identified as being important as virulence factors for a number of organ systems. Moreover, the virus attachment protein, σ1, of type 1 has been identified as being important for reovirus tropism to other organs as virus spreads from the intestine via the blood, whereas the reovirus type 3 σ1 protein targets neuronal cells in the gut and is associated with neuronal spread to the central nervous system following gut entry (TYLER et al. 1989).

7 Liver as a Site of Viral Clearance

Mesenteric veins drain the intestine and empty into the portal vein of the liver, where intestinally absorbed reovirus is cleared. This can be inferred indirectly from the finding that the liver has evidence of infection within 18h of peroral inoculation of 10^5 PFU of reovirus type 1 in adult mice, whereas other organs do not have measurable virus this early (RUBIN, unpublished observation). More direct evidence shows that when adult rats were injected intravenously with radiolabeled virus, within 5min over 90% of the virus had been cleared from the bloodstream. While the lungs cleared a disproportionate amount of reovirus type 1, the liver and spleen both accumulated high levels of both reovirus type 1 and 3 (VERDIN et al. 1987). These data indicate that both reovirus types 1 and 3 can be cleared from the blood by cells in the liver. Whether this reflects specific binding of reovirus to liver endothelial or tissue fixed macrophages (Kupffer cells) is uncertain.

Some of the virus that enters the liver by way of the bloodstream appears to be specifically exported into the bile. When the luminal contents of the intestines of mice intravenously infected with reovirus T1/L were titered 24h postinfection, it was found that the highest titers came from the duodenum. Since the duodenum is proximal to the ileum, the region of the intestine that is most intensely infected with T1/L, it was felt that virus was entering the lumen of the intestine from sites other than productively infected intestinal cells. In fact, when the bile duct was cannulated prior to infection and the bile was collected and titered at several time points postinfection, it was found that virus appeared in high titers in the bile as soon as 30min after infection, much too soon to be a result of newly replicated virus. Moreover, since the titers achieved in the bile are greater than the titers achieved in the blood and since infection with T3/D does not result in measurable titers in the bile, it is apparent that some form of specific active transport of T1/L from the circulatory system into the bile is at work (RUBIN et al. 1987).

This specific elimination of the T1/L virus from the hepatic system into the bowel seems to be driven by at least a subset of the Kupffer cells. These cells are known to selectively transport a macromolecule, immunoglobulin A (IgA), from systemic circulation and present it to hepatocytes for secretion into the bile. Several lines of evidence are consistent with Kupffer cells also transporting T1/L reovirus in a similar manner. First, electron microscopy has visualized T1/L virus within Kupffer cells as soon as 2h after intravenous inoculation, showing that the virus is internalized in these cells. Second, two agents which have been previously shown to inhibit uptake in Kupffer cells, silica dioxide and carrageenan, block the secretion of virus into the bile if they are administered to the mice prior to virus inoculation. Finally, highly purified IgG2a anti-I-A^K antibodies, which should bind to and inhibit the function of the 40%–60% of Kupffer cells that are positive for the I-A antigen, drastically reduce the amount of virus found in the bile while control IgG2a antibodies have no discernible effect (RUBIN et al. 1987).

Whether this selective transport of the virus from the liver has evolved in favor of the host or the virus is unknown. Does it primarily favor the host by exporting the infectious virus away from a sensitive organ? Or, does it primarily favor the virus by shunting it away from the closed systemic circulation to the bowel, an organ that ultimately releases it to the outside environment?

8 Liver as a Site of Infection

Evidence that liver cells not only transport virus into the biliary system but are also a potential site of reovirus infection has been obtained from adult severe combined immunodeficient (SCID) mice. When adult SCID mice were inoculated either orally or intraperitoneally with 10^5–10^8 PFU of reovirus type 1 or 3, after 2–5 weeks virus titers were obtained from all examined tissues. However, it was only within the liver that virus-induced lesions were found (RUBIN et al. 1990; HALLER et al. 1995a). The lesions were evident by the second week after infection and ranged from hepatocyte swelling with loss of cytoplasmic basophilia to frank eosinophilic bodies with pyknotic nuclei. There were also infiltrates of mononuclear and polymorphonuclear cells and areas of coagulation necrosis rimmed by an inflammatory infiltrate. The number of lesions increased in an infected mouse over time (RUBIN et al. 1990). With intraperitoneal inoculation of reovirus type 1 the mice were all dead by the end of 4 weeks while the mice orally inoculated with reovirus type 1 lived as long as 6 weeks (HALLER et al. 1995b; RUBIN et al. 1990).

While it is not possible to directly assign a cause of death, it was clearly associated with a severe hepatitis as necrotic tissue was only found within the liver. Moreover, while some reovirus antigen was found in intestinal crypt epithelial cells and Purkinje cells, it was most prevalent in the liver in hepatocytes and Kupffer cells surrounding areas of necrosis (RUBIN et al. 1990). These data are consistent with a model of infection in which only a few dividing liver cells are initially capable of replicating virus, and that the failure of the immune system to clear infection allows additional cells to enter the cell cycle in the presence of virus. Additionally, liver cells are recruited into cellular division by the loss of some cells due to virus infection.

9 Cellular Proliferation Increases Susceptibility of Liver Cells to Infection: In Vivo

The model of reovirus infection of dividing liver cells proposed in the setting of the murine SCID host that lacks anti-reovirus B- and T-cell responses has been substantiated. Reovirus type 1 induces lethal disease in SCID mice with extensive hepatocellular pathology, yet in adult immunocompetent mice reovirus type 1 does not target the liver. When adult mice were inoculated intravenously with approximately 10^9 PFU of reovirus type 1/L only about 1% of the hepatocytes were found to express reovirus antigen at the peak of infection (48h). By the fifth day after infection the virus had cleared completely with no evident morbidity or mortality (RUBIN et al. 1990). However, susceptibility of Kupffer cells and hepatocytes to reovirus type 1 infection can be increased dramatically by affecting the rate of cellular proliferation, such as by injury or hepatotoxins.

Both carbon tetrachloride (CCl_4) and α-naphthylisothiocyanate (ANIT) transiently damage the liver in a site preferential manner which stimulates regenerative cellular proliferation. CCl_4 causes centrizonal necrosis and subsequent regeneration while ANIT causes pericholangitis and pericholangiolitis with resultant inflammation and/or cellular proliferation near the portal tract. When either one of these toxins was administered to mice which were then inoculated via the tail vein with reovirus type 1/L, the amount of reovirus antigen present in the liver sections was higher than that in the liver sections obtained from mice given the virus alone.

Reovirus antigen expression in mice given CCl_4 was 10–20 times higher than control, and antigen was found in both hepatocytes and Kupffer cells adjacent to the areas of CCl_4 induced centrizonal necrosis. The mice given ANIT had a significantly increased amount of antigen evident in hepatocytes found within two cell diameters of the portal tract. Thus the pattern of increase of antigen expression in the two experiments reflected the known sites of regenerative cellular proliferation induced by the toxins. The increase in viral titers in the livers of the CCl_4-treated animals were also markedly higher than controls, reflecting the widespread damage caused by the toxin and the markedly higher antigen expression. In the ANIT-treated animals the increase in viral titers derived from whole livers was not statistically significant, consistent with the very local pattern of damage and increased antigen expression (Rubin et al. 1990; Piccoli et al. 1990).

There are two lines of evidence suggesting that the increased reovirus antigen expression and titers of virus in the livers of mice treated with CCl_4 is due to an increase in liver cell proliferation. First, the toxin-induced damage increased reovirus antigen expression in the liver compared to controls when the toxin was given simultaneously with the virus, or 2 or even 4 days before the virus. This is consistent with the known effect of the toxin to induce mitosis as soon as 12h after administration and peaking at 48–72h after administration. It is unlikely that some other effect, such as alterations in membrane fluidity caused by CCl_4 would lead to increases in virus infectivity for virus given as late as 4 days after hepatic exposure to the toxin. Second, there was a significant mortality in mice given both virus and CCl_4 but only when the toxin was given 1 day prior or simultaneous with the virus. That is, the synergistic effect of virus and toxin to produce death occurred at the same time as the expected peak of regenerative mitosis (Piccoli et al. 1990; Rubin et al. 1990).

Increased antigen production was also observed in livers that were stimulated to regenerate by physically induced trauma. Mice were surgically opened and then subjected to either a 50% hepatic resection, a blunt trauma to one lobe (pinched by forceps), or a sham operation (control). In these mice an increase in antigen expression was seen over control mice in the surgically damaged (regenerating) lobe. The other lobes of the liver did not show a statistically significant increase in virus antigen production. Strikingly, in the lobes that were traumatized by being pinched by forceps the increased antigen expression occurred discretely within the arch of wounded, regenerating tissue. (Piccoli et al. 1990; Rubin et al. 1990). These data are consistent with the preferential reovirus growth in proliferating cells, but in this case there is no danger of an artifact due to exogenous toxins. These studies demonstrate how innocuous viruses may develop into true pathogens with alternations in environmental exposures or the immune status of the host.

10 Cellular Proliferation Increases Susceptibility of Liver Cells to Infection: In Vitro

This selective infection of cycling hepatic cells by reovirus can also be demonstrated *in vitro* with a murine hepatocarcinoma cell line, Hepa 1/A1. Hepa 1/A1 cells were derived from a spontaneously arising murine hepatocarcinoma and are known to decrease the rate of cell cycling when becoming contact inhibited or on exposure to dimethylsulfoxide (DMSO). This is in contrast to L cells, which show no contact inhibition or decrease in cell cycling when exposed to DMSO. Since both L cells and Hepa 1/A1 cells have been shown to have similar numbers of receptors for reovirus, and when infected, produce similar titers of virus particles per cell, they together offer a novel model system in which to examine the propensity of reovirus to infect dividing vs. quiescent cells. First, only 10% of Hepa 1/A1 cells are infected when confluent monolayers are exposed to reovirus with a multiplicity of infection of 5. This inocula of virus is sufficient to infected nearly 100% of L cells under similar culture conditions. However, the number of infected Hepa 1/A1 cells increases over time, suggesting that the Hepa 1/A1 cell monolayers are able to replicate virus, perhaps due to cell infected cell death and recruitment of additional cells to divide into newly created spaces. Second, it was found that DMSO, which reduces cell cycling in Hepa 1/A1, is capable of diminishing the number of Hepa 1/A1 cells that can be infected. DMSO does not affect the capacity of reovirus to infect L cells nor L cell cycling. Third, it was found that in contact inhibited Hepa 1/A1 cells about 18% of the cells were actively undergoing cell cycling and strikingly, when these monolayers were infected about the same number (18%) became infected as measured by infectious center assay.

All of these findings point to a strong link between cell cycling in hepatic cells and ability to host reovirus growth. Moreover, when Hepa 1/A1 cells were arrested in a replicative state by the addition of aphidicolin, infected with reovirus, given 5'-bromo-2'-deoxyuridine (BrdU; which incorporates in the DNA of actively growing cells) and then cell-sorted based on BrdU incorporation, it was found by immunohistochemistry that there was a strong correlation between BrdU incorporation and virus growth. This indicates a tendency of reovirus for growth in cells actively replicating DNA (Taterka et al. 1994). It is also striking how similar this result is to the result obtained in vitro with intestinal cells where one can scratch a monolayer of cells and then observe preferential growth in the cells that grow into the wound in the monolayer (Organ et al., submitted).

11 Biliary Atresia: Experimental Models and Viral Determinants

In some of the earliest work with type 3 reovirus it was observed that inoculation of newborn mice with high titers led to what has come to be called "oily hair" syndrome. Subsequent work has associated this syndrome with damage to and occlusion of the biliary system. In fact, light and electron microscopy of neonatal mice infected with the LD_{50} of type 3 reovirus has shown that the majority of the epithelial cells lining the common bile duct were infected. This led to extensive swelling, edema and necrotic obstruction within the hepatobiliary tree (TYLER and FIELDS 1996; PAPADIMITRIOU 1968). It has been suggested that the oily hair syndrome of these mice is a result of this kind of obstruction to the biliary systems which results in a loss of bile flow to the intestine with the resultant malabsorption of fats.

This extensive damage to the biliary system has been clearly demonstrated to be related to tropism of some strains of type 3 reovirus for the biliary system rather than due to simple virulence. A battery of different strains of type 3 virus were tested on neonatal mice with oral inoculation to assay for their LD_{50} titers and ability to cause the oily hair syndrome. Of the seven tested strains only two, T3C31 and T3 Abney, caused oily hair syndrome. T3C31, had an LD_{50} at greater than 10^8 PFU per mouse while T3 Abney, had an LD_{50} at 2×10^2 PFU. Both of these strains were able to cause oily hair with a PFU dosage more than an order of magnitude below their LD_{50} dosage. While the other strains had an equally great range of LD_{50} values, they did not cause oily hair at even the LD_{50} dosage. Not only was the virulence of the virus unrelated to the ability to cause oily hair, the ability of the virus to grow in the liver was not correlated with oily hair either.

When the growth in the liver was titered for T3 Abney, which causes oily hair and T1/L and T3C43 which does not, it was found that all had grown to equal titers in infected livers by the tenth day after inoculation. However, only the T3 Abney infected mice developed extensive hepatic damage and occlusions of the bile ducts. Finally, immunoperoxidase staining of liver sections in these animals indicated the presence of reovirus antigen in the areas of damage in the T3 Abney infected animals, including the bile duct epithelia and the area surrounding the bile ducts. The T1/L- and T3C31-infected animals had a more diffuse pattern of hepatic staining (WILSON et al. 1994). Together these data show that the subtle differences among the reovirus strains can result in dramatic differences in biliary tropism and pathogenesis.

Using reassortants between T3 Abney and T1/L, it was found that only the S1 gene segment is correlated with the ability to cause biliary disease (WILSON et al. 1994). While this finding differs from that of a previous study that was unable to map the biliary disease causing ability to a single virus gene segment (BANGARU et al. 1980), it is not possible to compare the results directly as there were numerous differences in methodologies. When the sequences of the S1 gene of the two disease causing strains, T3 Abney and T3C31, were compared with the S1 gene sequences of nine other T3 isolates that do not cause biliary disease it was found that both T3

Abney and T3C31 have amino acid substitutions not found in any of the other strains. In fact, each had a unique substitution within eight residues of each other within an otherwise highly conserved region of the cell attachment domain (WILSON et al. 1994). This suggests that this highly conserved region of the attachment domain of S1 determines the unique tropism for some type 3 strains for the biliary duct epithelium.

The tropism of reovirus type 3 has also been shown in one set of experiments to extend to transplacental transmission. First, reovirus type T3/D was passaged twice through neonatal mice livers to increase its hepatotropism. Then pregnant mice were inoculated with it, and the progeny were assayed for the presence of reovirus RNA by dot-blot hybridization and PCR. Not only was the RNA present in both the embryonic and fetal progeny, but it was also present in the neonatal mice as late as 3 weeks after birth. Curiously, the progeny failed to demonstrate any clinical symptoms of disease, although, there was some histological evidence of infection in the liver and bile ducts. Clearly the virus is able to cross over from the mother to the offspring and infect the developing hepatic system. Moreover, the persistence of the virus indicated by the lingering presence of reovirus RNA was also demonstrated in mice that had been injected as newborns. They turned out to have the reovirus RNA in their livers as long as 7 weeks after injection (PARASHAR et al. 1992). This is a significant finding because there is some evidence that the ability of a virus to persist in a developing or immune deficient host can lead to disease with virus strains that are otherwise avirulent. For example, the T3C9 strain, which does not normally cause oily hair (WILSON et al. 1994) causes oily hair in mice depleted of $CD4^+$ and $CD8^+$ cells (VIRGIN and TYLER 1991). This raises the distinct possibility that an infection with reovirus in the embryonic stage allows for the development of virus mutants which differ in tropism perhaps due to unrestricted replication. Alternatively, maternal antibody provides a selective pressure for variants to emerge. Further work is needed to define the mechanism by which variants of reovirus arise under fetal conditions.

12 Biliary Atresia: Human is Correlated to Experimental Models

A controversy has developed regarding whether extrahepatic biliary atresia and neonatal hepatitis in human neonatal infants is associated with a type 3 reovirus infection. It has long been suspected that these diseases are caused by an unidentified perinatal viral infection. The similarities of their hepatobiliary pathologies to those of mice with oily hair syndrome has led some authors to suspect that reovirus type 3 infection is involved in the human disease (TYLER and FIELDS 1996). A series of papers has been published reporting a correlation between human infants with extrahepatic biliary atresia or neonatal hepatitis and the presence of anti-reovirus antibodies in the serum (BANGARU et al. 1980; MORECKI et al. 1982; GLASER et al. 1984). Reovirus antigen and viruslike particles were also found in diseased tissue from an infant (MORECKI et al. 1984). It has even been reported that a Rhesus monkey with biliary atresia was found to have serum positive for anti-reovirus antibodies (ROSENBERG et al. 1983). Unfortunately, other studies have not found a correlation between perinatal reovirus infection and human biliary atresia or neonatal hepatitis (PARASHAR et al. 1992; WILSON et al. 1994; DUSSAIX et al. 1984; BROWN et al. 1987). Moreover, a recent study to use PCR to find evidence of reovirus RNA in samples of hepatobiliary tissues obtained from diseased infants was unsuccessful (STEELE et al. 1995). Therefore a role of reovirus type 3 in the development of neonatal liver disease or biliary atresia has not been clearly established.

13 Summary and Future Directions

While the involvement of reovirus with human disease is still open, the insights gained on the capacity of reoviruses to enter the host and passage to the central nervous system and systemic organs have opened new vistas. The capacity of reovirus selectively to replicate within the liver, biliary, and gastrointestinal system as demonstrated by experimental reovirus infection provides new insight into the role of cellular factors that participate in the establishment of pathogenic infections. Further work is in progress to identify cellular genes associated with reovirus replication. Finally, the unique tropism for reovirus to selected sites with the intestine and hepatobiliary system indicates how a virus may be associated with diseases such as intestinal and biliary atresia, and suggests that further search for an infectious origin may be fruitful. Further work may resolve whether some cases of liver disease in neonates or patients with the acquired immunodeficiency syndrome may be due to the reovirus.

Acknowledgements. We thank E. Eisenberg and L. Organ for their review of the manuscript. This work was supported by grants from the National Cancer Institute (CA68283, DHR).

References

Ambler L, MacKay M (1991) Reovirus1 and 3 bind and internalize at the apical surface of intestinal epithelial cells. Virology 18:162–169

Amerongen HM, Wilson GAR, Fields BN, Neutra MR (1994) Proteolytic processing of reovirus is required for adherence to intestinal M cells. J Virol 68:8428–8432

Bangaru B, Morecki R, Glaser JH, Gartner LM, Horwitz MS (1980) Comparative studies of biliary atresia in the human newborn and reovirus-induced cholangitis in weanling mice. Lab Invest 43:456–462

Bass DM, Trier JS, Dambrauskas R, Wolf RJ (1988) Reovirus type 1 infection of small intestinal epithelium in suckling mice and its effect on M cells. Lab Invest 58:226–235

Bass DM, Bodkin D, Dambrauskas R, Tier JS, Fields BN, Wolf JL (1990) Intraluminal proteolytic activation plays an important role in replication of type 1 reovirus in the intestines of neonatal mice. J Virol 64:1830–1833

Blay J, Brown KD (1984) Characterization of an epithelioid cell line derived from rat small intestine: demonstration of cytokeratin filaments. Cell Biol Int Rep 8:551–560

Blay J, Brown KD (1985) Epidermal growth factor promotes the chemotactic migration of cultured rat intestinal epithelial cells. J Cell Phys 124:107–112

Bodkin DK, Fields BN (1989) Growth and survival of reovirus in intestinal tissue: role of the L2 and S1 genes. J Virol 63:1188–1193

Bodkin DK, Nibert ML, Fields BN (1989) Proteolytic digestion of reovirus in the intestinal lumens of neonatal mice. J Virol 63:4676–4681

Branski D, Lebenthal E, Faden H, Hatch TF, Krasner J (1980) Small intestinal epithelial brush border enzymatic changes in suckling mice infected with reovirus type 3. Pediatr Res 14:803–805

Brown WR, Sokol RJ, Levin MJ, Silverman A, Tamaru T, Lilly JR, Hall RJ, Cheney M (1987) Lack of correlation between infection with reovirus 3 and extrahepatic biliary atresia or neonatal hepatitis. J Pediatr 113:670–676

Dryden KA, Wang G, Yeager M, Nibert ML, Coombs KM, Furlong DB, Fields BN, Baker TS (1993) Early steps in reovirus infection are associated with dramatic changes in supramolecular structure and protein conformation: analysis of virons and subviral particles by cryoelectron microscopy and image reconstruction. J Cell Biol 122:1023–1041

Duncan MR, Stanish SM, Cox DC (1978) Differential sensitivity of normal and transformed human cells to reovirus infection. J Virol 28:444–449

Dussaix E, Hadchouel M, Tardieu M, Alagille D (1984) Biliary atresia and reovirus type 3 infection. N Engl J Med 310:658

Glaser JH, Balistreri WF, Morecki R (1984) Role of reovirus type 3 in persistent infantile cholestasis. J Pediatr 105:912–915

Haller BL, Barkon ML, Vogler GP, Virgin HW IV (1995a) Genetic mapping of reovirus virulence and organ tropism in severe combined immunodeficient mice: organ specific virulence genes. J Virol 69:357–364

Haller BL, Barkon ML, Li XY, Hu WH, Wetzel JD, Dermody TS, Virgin HW IV (1995b) Brain-and intestine-specific variants of reovirus serotype 3/D are selected during chronic infection of severe combined immunodeficient mice. J Virol 69:3933–3937

Hrdy DB, Rosen L, Fields BN (1979) Polymorphism of the migration of double-stranded RNA genome segments of reovirus isolates from humans, cattle, and mice. J Virol 31:104–111

Jackson GG, Muldoon RL, Cooper RS (1961) Reovirus type 1 as an etiologic agent of the common cold. J Clin Invest 40:1051

Joklik WK (1972) Studies on the effect of chymotrypsin on reovirus. Virol 49:700–801

Kauffman RS, Wolf JL, Finberg R, Trier JS, Fields BN (1983) The sigma 1 protein determines the extent of spread of reovirus from the gastrointestinal tract of mice. Virology 124:403–410

Keroack M, Fields BN (1986) Viral shedding and transmission between hosts determined by reovirus L2 gene. Science 232:1635–1638

Leers WD, Royce KR (1966) A survey of reovirus antibodies in sera of urban children. Can Med Assoc J 94:1040–1042

Lerner AM, Cherry JD, Klein JO, Finland M (1947) Infections with reoviruses. N Engl J Med 267:947–952

Morecki R, Glaser JH, Cho S, Balistreri WF, Horwitz MS (1982) Biliary atresia and reovirus type 3 infection. N Engl J Med 307:481–484

Morecki R, Glaser JH, Johnson AB, Kress Y (1984) Detection of reovirus type 3 in the porta hepatis of an infant with extrahepatic biliary atresia: ultrastructural and immunocytochemical study. Hepatology 4:1137–1142

Morrison LA, Sidman RL, Fields BN (1991) Direct spread of reovirus from the intestinal lumen to the central nervous system through the vagal autonomic nerve fibers. Proc Natl Acad Sci USA 88:3852–3856

Nibert ML, Furlong DB, Fields BN (1991) Mechanisms of viral pathogenesis: distinct forms of reoviruses and their roles during replication in cells and host. J Clin Invest 88:727–734

Nibert ML, Chappel JD, Dermody TS (1995) Infectious subvirion particles of reovirus type 3 Dearing exhibit a loss in infectivity and contain a cleaved sigma 1 protein. J Virol 69:5057–5067

Nibert ML, Schiff LA, Fields BN (1996) Reoviruses and their replication. In: Fields BN, Knipe DM, Howley PM et al (eds) Fields virology, 3rd edn. Lippincott-Raven, Philadelphia, pp 1557–1596

Organ EL, Nalbantyan CD, Nanney LB, DuBois RN, Coffey RJ, Rubin DH (submitted) In vitro modeling of reovirus infection utilizing rat intestinal epithelial cells (RIE-1): transforming growth factor alpha (TGF-alpha) and insulin-like growth factor increase viral replication in the presence of serum

Papadimitriou JM (1968) The biliary tract in acute murine reovirus 3 infection. Am J Pathol 52:595–601
Parashar K, Tarlow MJ, McCrae MA (1992) Experimental reovirus type 3-induced murine biliary tract disease. J Pediatr Surg 27:843–847
Piccoli DA, Witzleben CL, Guico CJ, Morrison A, Rubin DH (1990) Synergism between hepatic injuries and a nonhepatotropic reovirus in mice. J Clin Invest 86:1038–1045
Reinhardt MC (1984) Macromolecular absorption of food antigens in health and disease. Ann Allergy 53:597–601
Rosen L, Abinanti FR (1960) Natural and experimental infection of cattle with human types reoviruses. Am J Hyg 71:250–257
Rosenberg DP, Morecki R, Lollini LO, Glaser J, Cornelius CE (1983) extrahepatic biliary atresia in a Rhesus monkey (Maraca mulatta). Hepatology 3:577–580
Rubin DH (1987) Reovirus serotype 1 binds to the basolateral membrane of intestinal epithelial cells. Microb Pathog 3:215–219
Rubin DH, Fields BN (1980) Molecular basis of reovirus virulence: role of the M2 gene. J Exp Med 152:853–868
Rubin DH, Kornstein MJ, Anderson AO (1985) Reovirus serotype 1 intestinal infection: a novel replicative cycle with ileal disease. J Virol 53:391–398
Rubin DH, Eaton MA, Costello T (1986a) Reovirus type 1 is secreted into the bile. J Virol 60:726–728
Rubin DH, Eaton MA, Anderson AO (1986b) Reovirus infection in adult mice: the virus hemagglutinin determines the site of intestinal disease. Microb Pathog 1:79–87
Rubin DH, Costello T, Witzleben CL, Greene MI (1987) Transport of infectious reovirus into bile: class II major histocompatibility antigen-bearing cells determine reovirus transport. J Virol 61:3222–3226
Rubin DH, Morrison AH, Witzleben CL, Guico CJ, Piccoli DA (1990) Site of reovirus replication in liver is determined by the type of hepatocellular insult. J Virol 64:4593–4597
Shiner M (1985) Microscopic anatomy of the small intestine. In: Beck JE (ed) Gastroenterology. Saunders, Philadelphia, pp 1479–1491
Spendlove RS, McClain ME, Lennette EH (1970) Enhancement of reovirus infectivity by extracellular removal or alteration of the virus capsid by proteolytic enzymes. J Gen Virol 8:83–94
Steele MI, Marshall CM, Lloyd RE, Randolph VE (1995) Reovirus 3 not detected by reverse transcriptase mediated polymerase chain reaction analysis of preserved tissue from infants with cholestatic liver disease. Hepatology 21:697–702
Taterka J, Sutcliffe MS, Rubin DH (1994) Selective reovirus infection of murine hepatocarcinoma cells during cell division. J Clin Invest 94:353–360
Tyler KL, Fields BN (1996) Reoviruses. In: Fields BN, Knipe DM, Howley PM et al (eds) Fields virology, 3rd edn. Lippincott-Raven, Philadelphia, pp 1597–1623
Tyler KL, Virgin HW 4th, Bassel-Duby R, Fields BN (1989) Antibody inhibits defined stages in the pathogenesis of reovirus serotype 3 infection of the central nervous system. J Exp Med 170 (3):887–900
Verdin EM, Maratos-Flier E, Kahn CR, Sodoyez JC, Sodoyez-Goffaux F, De Vos CJ, Lynn SP, Fields BN (1987) Visualization of viral clearance in the living animal. Science 236:439–442
Virgin HW IV, Tyler KL (1991) Role of immune cells in protection against and control of reovirus infection in neonatal mice. J Virol 65:5157–5164
Walker WA (1979) Gastrointestinal host defense: importance of gut closure in control of macromolecular transport. Ciba Found Symp 70:201–219
Weiner DB, Girard K, Williams WV, McPhillips T, Rubin DH (1988) Reovirus type 1 and type 3 differ in their binding to isolated intestinal epithelial cells. Microb Pathog 5:25–40
Wilson GAR, Morrison LA, Fields BN (1994) Association of the reovirus S1 gene with serotype3-induced biliary atresia in mice. J Virol 68:6458–6465
Wolf JL, Rubin DH, Finberg R, Kauffman RS, Sharpe AH, Trier JS, Fields BN (1981) Intestinal M cells: a pathway for entry of reovirus into the host. Science 212:471–472
Wolf JL, Kauffman RS, Finberg R, Dambrauskas R, Fields BN, Trier JS (1983) Determinants of reovirus interaction with intestinal M cells and absorptive cells of murine intestine. Gastroenterology 85:291–300
Wolf JL, Dambrauskas R, Sharpe AH, Trier JS (1987) Adherence to and penetration of the intestinal epithelium by reovirus type 1 in neonatal mice. Gastroenterology 92:82–91

Reovirus and Endocrine Cells

E. MARATOS-FLIER

1 Introduction

Early studies on the effects of reovirus infection of mice noted the development of autoimmune disease associated with a runting syndrome (STANLEY and WALTERS 1966). Infection of endocrine tissues was not initially reported, and runting was attributed directly to autoimmunity rather than to endocrine dysfunction.

2 Animal Studies: Pancreas and Pituitary

The potential interaction of reoviruses with endocrine cells was first examined by ONODERA et al. (1978), pursuing the hypothesis that viruses act as etiological agents in the development of diabetes. The diabetogenic effects encephalomyocarditis virus (NOTKINS 1977) in genetically susceptible (ROSS et al. 1976; BOUCHER et al. 1975) strains of mice had been reported. These initial studies examined the effects of reovirus type 3 passaged in primary β-cell cultures on neonatal mice. Animals were infected with 1.0×10^5PFU virus. Viral infection led to a fall of pancreatic insulin

Research Division, Joslin Diabetes Center, and Department of Medicine, Harvard Medical School, Joslin Diabetes Center, 1 Joslin Place, Boston MA 02215, USA
(E-mail: emarat@joslab.harvard.edu)

levels that persisted for a 2-week period and was associated with abnormal glucose tolerance. Histological examination of the pancreas performed 5 days after infection revealed coagulation necrosis of some of the islets. Electron microscopy revealed the presence of reticulogranular viral matrix characteristic of reovirus. Using double antibody labeling, it was possible to demonstrate the presence of reovirus antigens in β-cells.

Subsequent studies with type 3 reovirus demonstrated viral growth in cultures of human β-cells and enhancement of growth by repeated passage through these cultures (Yoon et al. 1981). Insulin content of infected cells was significantly lower than content of uninfected cells. Similarly, secretion of insulin into the culture medium was reduced.

A more extensive study examined the endocrine effects of infection with type 1 reovirus (Onodera 1981). Infected mice demonstrated impaired glucose tolerance and a decrease in pancreatic immunoreactive insulin. On average 20% of cells in each islet were positive for viral antigens. Electron microscopy revealed that viral particles were seen primarily in β-cells, however α- and δ-cells were also infected. Significant numbers of mice exhibited growth retardation, suggesting that the pituitary gland is also a target of reovirus type 1. Examination of the pituitary by electron microscopy revealed the presence of viral particles in growth hormone producing cells. Mice exhibiting growth retardation had on average a more than 50% reduction in circulating plasma growth hormone levels.

Since reovirus infections are associated with autoimmunity, the possibility that endocrinopathies, seen in type 1 infected mice, are mediated by autoantigens was investigated. Using indirect immunofluorescence, autoantibodies against anterior pituitary, islets, and gastric mucosa were detected. Antibodies against adrenal, thyroid, ovary, or testes were not found. Induction of autoantibodies was age dependent; 100% of neonatal mice infected at 3 days of age developed autoantibodies to growth hormone; mice infected at 30 days of age failed to develop antibodies.

The development of autoimmunity appeared to be specific to the type 1 serotype. Type 3 infection did not induce autoantibodies. Using viral reassortants it was possible to show that the ability to induce autoantibodies to growth hormone was a viral characteristic that could be mapped to the S1 gene, which codes for the viral hemagglutinin and which determines viral tropism. The reovirus type 1 S1 gene targets virus to the anterior pituitary as well as ependyma. Presumably growth of virus in the pituitary is required (but not sufficient) for development of antibodies; however, the molecular mechanism by which reovirus induces autoantibodies is still unknown. Reovirus type 3 is capable of replicating in β-cells but fails to induce autoantibodies.

Immunosuppression alters the host response to reovirus infection (Onodera et al. 1982). Mice were infected with type 1 reovirus and either rabbit anti-mouse thymocyte serum or rabbit anti-mouse lymphocyte serum or cyclophosphamide (20mg/kg) were used for immunosuppression. Immunosuppressed mice did not develop glucose intolerance. Development of both autoantibodies and the runting syndrome could be suppressed by treatment of mice with antilymphocyte serum.

Immunosuppression did not alter the pattern of viral growth in liver, heart, or pancreas and reduced but did not eliminate the antibody response to virus.

Spleen cells from infected mice that developed antibodies could be used to make hybridomas that secreted monoclonal antibodies to endocrine tissues (HASPEL et al. 1983). Monoclonal antibodies reacted against pancreas, pituitary, stomach, and cell nuclei. Thirteen hybridomas made antibodies against islets. These antibodies tended to react with cells at the periphery of the islet. Only one hybridoma yielded antibodies reactive with the central portion of the islet where the β-cells are located. This antibody had high activity against rat insulin. Similarly, some of the antibodies directed against pituitary reacted with epitopes on the growth hormone molecule.

A recent study demonstrated that mice infected with type 2 reovirus also develop diabetes as an immune-mediated syndrome (HAYASHI et al. 1995).

3 Thyroid

Although initial studies failed to detect a thyroid autoimmune response, reovirus can cause autoimmune thyroiditis. Both SJL/J and BALB/c mice infected with T1 reovirus developed thyroiditis associated with autoantibodies to thyroglobulin (ONODERA and AWAYA 1990; SRINIVASAPPA et al. 1988). Histologically an infiltration of mononuclear cells and focal thyroiditis is seen, although there are no changes in thyroid function. Treatment with synthetic thymic factor suppressed the antibody response. T3 reovirus was unable to infect the thyroid in the mouse, and an autoimmune response was not elicited. The differential ability of T1 and T3 reovirus to infect the thyroid was mapped to the viral hemagglutinin (S1 gene product), suggesting that the hemagglutinin mediates viral targeting to this tissue.

4 In Vitro Studies

In vivo the ability of reovirus to induce an autoimmune response is dependent on the ability of virus to target and grow in endocrine tissues. Spread after an intramuscular injection is determined by the viral hemagglutinin (TYLER et al. 1986) as is clearance from the blood (VERDIN et al. 1988). In tissue culture successful viral infection requires a series of interactions between the virus and the host cell and is dependent on intact cellular function. The different effects of type 1 and type 3 reovirus on pituitary and islet tissue in vivo reflect the result of interactions of the virus with both the host organism and host tissue. Studies in vitro make it possible to dissect out effects that are specific to virus host-cell interaction.

Type 1 and type 3 reovirus can infect pituitary (GH_4C_1) and islet (RINm5F) derived cells and growth curves of both the viral serotypes in both tissues are remarkably similar (MARATOS FLIER et al. 1985). The virus enters its eclipse phase within 2h, and maximal titers of about 1×10^7 PFU per 5×10^5 cells are seen within 24h. These similar growth curves are seen despite the fact that the receptors for type 1 and type 3 reovirus are distinct (EPSTEIN et al. 1984; MARATOS FLIER et al. 1983) and indicate that receptor number and virus-receptor affinity as well as receptor mediated uptake and processing are essentially the same for both viruses in both cell types.

In this context it is noteworthy that the cellular responses to infection vary. Infection of RINm5F with either T1 or T3 reovirus is relatively cytotoxic, and approximately 50% of infected cells are killed 3 days after infection with a dose of 4 PFU per cell. T1 reovirus is similarly cytotoxic to GH_4C_1 cells, however T3 reovirus infection led to only 20% cell mortality, 3 days after infection. The ability of viruses to inhibit DNA synthesis was not a good predictor of cytotoxicity. T3, which was less effective at inhibiting DNA synthesis in RINm5F cells, was no less cytotoxic than T1 virus. In GH_4C_1 cells the ability of cells to synthesize DNA was only minimally impaired by T1 and T3 virus; however, T1 infection led to more cytotoxicity.

5 Reovirus and MHC

A potential mechanism by which reovirus infection might lead to autoimmunity was explored in a series of studies examining the effect of reovirus infection on expression of major histocompatibility complex (MHC) antigens (CAMPBELL and HARRISON 1989). RINm5F cells infected with either type 1 or type 3 reovirus showed up to a tenfold increase in class I MHC expression. The degree of increase was similar to that seen in cells treated with interferon-γ (100U/ml) for 24h. RINm5F cells do not express class II MHC and infection with type 3 reovirus failed to induce expression of this antigen. Expression of class I proteins required viral growth. UV irradiated virus was significantly less effective in inducing class I MHC expression. Induction of MHC expression was also seen in human β-cells infected with type 3 reovirus (CAMPBELL et al. 1988).

Although the effect of reovirus infection is similar to that of interferon treatment on induction of class I MHC expression, interferon and reovirus had different effects on induction of expression of intercellular adhesion molecule 1 which mediates cell-cell adhesion, an important factor in autoimmune responses. Treatment with interferon-γ induced ICAM-1 expression in human islets. Induction was seen on both insulin-secreting and somatostatin-secreting cells. However infection with type 3 reovirus had no effect on ICAM expression (CAMPBELL et al. 1989).

In thyroid cells reovirus infection leads to alterations in expression of both class I and class II (Ia) antigens. Type 1 reovirus induced de novo expression of class II antigens in the thyroid follicular epithelial cell line M.5 and augmented expression of class I antigens (GAULTON et al. 1989). Similar results were obtained using type 3 reovirus in a different thyroid cell line, 1B-6 (NEUFELD et al. 1989); class II antigens could also be induced, with a somewhat different time course, by interferon-γ.

Induction was seen even with UV-irradiated virus, indicating that viral binding was sufficient for stimulating induction and that viral replication was not required. Supernatants from infected cell cultures could also induce class II antigens. Supernatants pretreated with an anti-αβγ-interferon mix retained the ability to induce class II antigens, indicating that the antigen induction was not mediated through interferon. Supernatants pretreated with antiviral antibodies lost the ability to induce class II antigens. Somewhat different results were reported in human thyroid follicular cells (ATTA et al. 1995). Both T1 and T3 reovirus enhanced MHC class I expression on cells derived from eight different donors. Class II expression was strongly induced by type 1 reovirus in only 1 donor. Class II antigens could be weakly induced by either T1 or T3 reovirus in five other donors. Enhancement of class I antigens appears to be mediated through either α- or β-interferon as reovirus

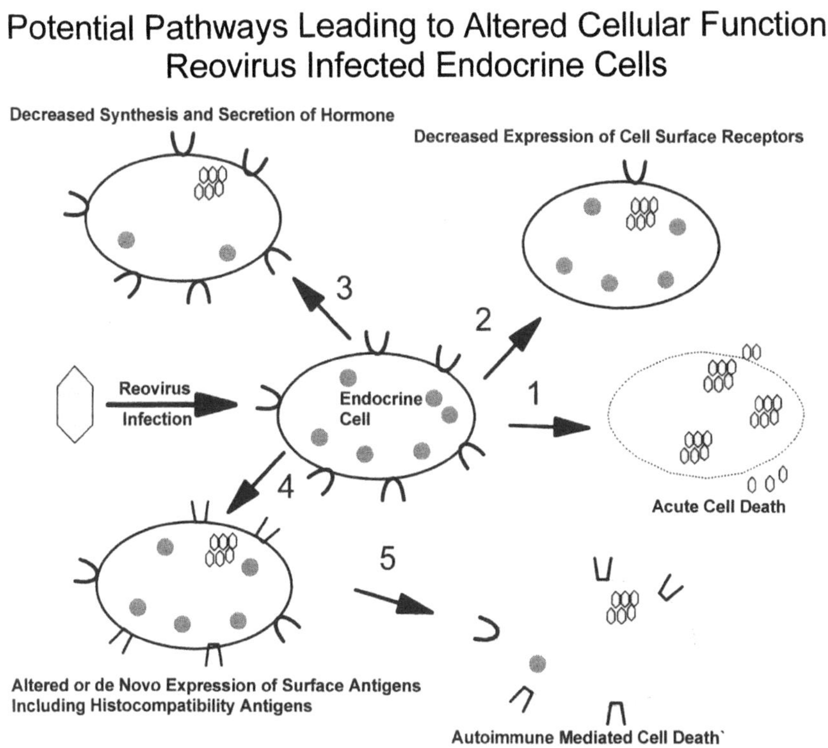

Fig. 1. An endocrine cell may be infected by reovirus. Possible outcomes include: rapid viral replication leading to acute cell death and lysis (1); persistent infection with changes in expressed membrane proteins but maintenance of normal levels of hormone synthesis (2); persistent infection with decreased hormone synthesis but normal expression of membrane proteins (3); persistent infection with altered or de novo expression of cell surface antigens and continued maintenance of normal hormone synthesis (4); autoimmune mediated cell death as a result of de novo expression of antigenically active proteins (5).

infected supernatants treated with antibodies of these interferons could not enhance class I expression.

While it is clear that reovirus infection alters MHC expression in endocrine cells the data are incomplete. Reovirus enhances class I antigens in RINm5F cultures but it fails to induce class II expression. It is unclear whether islets or primary islet cell cultures respond similarly, as infection of these tissues has not yet been comprehensively studied. In addition, the somewhat conflicting data derived from thyroid cell lines and primary thyroid cultures indicate that further studies of thyroid tissue are needed.

Still the ability of reovirus to enhance class I MHC expression and to induce class II expression is of particular interest in the context of the etiology of endocrine diseases such as type I diabetes and autoimmune thyroiditis. In the human pancreas, in early stages of type I diabetes, there is hyperexpression of class I MHC on all islet cells and a de novo expression of class II antigens specifically on β-cells (Foulis et al. 1987). Thyroid cells from patients with Grave's disease aberrantly express class II antigens (Bottazzo et al. 1983). These intriguing associations between reovirus and endocrine autoimmunity require further study.

Reovirus is known to cause persistent infection (Verdin et al. 1986), and it is possible that endocrinopathy results from persistent infection associated with aberrant cell surface antigen expression. In addition, persistently infected cells may demonstrate impaired responses to molecular signals secondary to altered receptor expression.

6 Clinical Correlations

Clinical correlations between reovirus infection and endocrinopathy in humans are quite limited. One study showed an increased prevalence of antibodies of to reovirus RNA as well as synthetic ds RNA in diabetics and their first-degree relatives (Huang et al. 1981). However, antibodies to ssDNA and ssRNA were also increased making it difficult to draw conclusions about potential role of reovirus infection in the etiology of human diabetes.

References

Atta MS, Irving WL, Powell RJ, Todd I (1995) Enhanced expression of MHC class I molecules on cultured human thyroid follicular cells infected with reovirus through induction of type 1 interferons. Clin Exp Immunol 101:121–126

Bottazzo GF, Pujol Borrell R, Hanafusa T, Feldmann M (1983) Role of aberrant HLA-DR expression and antigen presentation in induction of endocrine autoimmunity. Lancet 2:1115–1119

Boucher DW, Hayashi K, Rosenthal J, Notkins AL (1975) Virus induced diabetes mellitus. III. Influence of the sex and strain of the host. J Infect Dis 131:462–466

Campbell IL, Harrison LC (1989) Viruses and cytokines: evidence for multiple roles in pancreatic beta cell destruction in type 1 insulin dependent diabetes mellitus. J Cell Biochem 40:57–66
Campbell IL, Harrison LC, Ashcroft RG, Jack I (1988) Reovirus infection enhances expression of class I MHC proteins on human beta-cell and rat RINm5F cell. Diabetes 37:362–365
Campbell IL, Cutri A, Wilkinson D, Boyd AW, Harrison LC (1989) Intercellular adhesion molecule 1 is induced on isolated endocrine islet cells by cytokines but not by reovirus infection. Proc Natl Acad Sci USA 86:4282–4286
Epstein RL, Powers ML, Rogart RB, Weiner HL (1984) Binding of 125I labeled reovirus to cell surface receptors. Virology 133:46–55
Foulis AK, Farquharson MA, Hardman R (1987) Aberrant expression of class II major histocompatibility complex molecules by B cells and hyperexpression of class I major histocompatibility complex molecules by insulin containing islets in type 1 (insulin dependent) diabetes mellitus. Diabetologia 30:333–343
Gaulton GN, Stein ME, Safko B, Stadecker MJ (1989) Direct induction of Ia antigen on murine thyroid derived epithelial cells by reovirus. J Immunol 142:3821–3825
Haspel MV, Onodera T, Prabhakar BS, Horita M, Suzuki H, Notkins AL (1983) Virus induced autoimmunity: monoclonal antibodies that react with endocrine tissues. Science 220:304–306
Hayashi T, Yamamoto S, Onodera T (1995) Prevention of reovirus type 2 induced diabeteslike syndrome in DBA/1 suckling mice by treatment with antibodies against intercellular adhesion molecule and lymphocyte function associated antigen. Int J Exp Pathol 76:403–409
Huang SW, Haedt LH, Rich S, Barbosa J (1981) Prevalence of antibodies to nucleic acids in insulin-dependent diabetics and their relatives. Diabetes 30:873–874
Maratos Flier E, Spriggs DR, Fields BN, Kahn CR (1983) Specific plasma membrane receptor for reovirus on rat pituitary cells in culture. J Clin Invest 72:617–621
Maratos Flier E, Goodman MJ, Fields BN, Kahn CR (1985) Differential effects of viral infection on islet and pituitary cells in culture. Endocrinology 116:2430–2437
Neufeld DS, Platzer M, Davies TF (1989) Reovirus induction of MHC class II antigen in rat thyroid cells. Endocrinology 124:543–545
Notkins AL (1977) Virus induced diabetes mellitus. Arch Virol 54:1–17
Onodera T, Jenson AB, Yoon JW, Notkins AL (1978) Virus-induced diabetes mellitus: reovirus infection of pancreatic beta cells in mice. Science 201:529–531
Onodera T, Awaya A (1990) Antithyroglobulin antibodies induced with recombinant reovirus infection in BALB/c mice. Immunology 71:581–585
Onodera T, Toniolo A, Ray UR, Jenson AB, Knazak RA, Notkins AL (1981) Virus induced diabetes mellitus XX. Polyendocrinopathy and Autoimmunity J Exp Med 153:1457–1473
Onodera T, Ray UR, Melez KA, Suzuki H, Toniolo A, Notkins AL (1982) Virus induced diabetes mellitus: autoimmunity and polyendocrine disease prevented by immunosuppression. Nature 297:66–68
Ross ME, Onodera T, Brown KS, Notkins AL (1976) Virus-induced diabetes mellitus. IV. Genetic and environmental factors influencing the development of diabetes after infection with the M variant of encephalomyocarditis virus. Diabetes 25:190–197
Srinivasappa J, Garzelli C, Onodera T, Ray U, Notkins AL (1988) Virus induced thyroiditis. Endocrinology 122:563–566
Stanley NF, Walters NM (1966) Virus induction autoimmune disease and neoplasia. Lancet 1:962–963
Tyler KL, McPhee DA, Fields BN (1986) Distinct pathways of viral spread in the host determined by reovirus S1 gene segment. Science 233:770–774
Verdin EM, Maratos Flier E, Carpentier JL, Kahn CR (1986) Persistent infection with a nontransforming virus leads to altered expression of growth factor receptors and responses. J Cell Physiol 128:457–465
Verdin EM, Lynn.SP, Fields BN, Maratos Flier E (1988) Uptake of reovirus serotype I by the lungs from the bloodstream is mediated by the viral hemagglutinin. J Virol 62:545–551
Yoon JW, Selvaggio S, Onodera T, Wheeler J, Jenson AB (1981) Infection of cultured human pancreatic B cells with reovirus type 3. Diabetologia 20:462–467

Pathogenesis of Reovirus Infections of the Central Nervous System

K.L. TYLER

Departments of Neurology, Medicine, and Microbiology and Immunology, University of Colorado Health Sciences Center, 4200 East 9th Avenue, Denver, CO 80262, USA

Neurology Service (127), Denver Veterans Affairs Medical Center, 1055 Clermont Street, Denver, CO 80220, USA

1 Introduction

Reovirus infection in rodents, and predominantly mice, has been extensively utilized as an experimental model system for studying the pathogenesis of viral disease of the CNS (for review see TYLER and FIELDS 1988, 1996; TYLER 1991; VIRGIN et al. 1997). Rare cases of reovirus-induced neurological disease in humans, including encephalitis and meningitis (VAN TONGEREN 1957; KRAINER and ARONSON 1959; JOSKE et al. 1964; JOHANSSON et al. 1996; TYLER, unpublished) have been reported. Isolated case reports purporting to show an association between reovirus infection and atypical forms of motor neuron disease and chronic mental illness are totally unconvincing (AVERBACK 1982; SZIRMAI et al. 1983). Reovirus has also been associated with neurological illnesses in nonhuman animals, including hydrocephalus in monkeys (SABIN 1959), encephalitis in dogs (MASSIE and SHAW 1966), and ataxia in cats (CSIZÁ 1974).

In order for reoviruses to infect and injure the central nervous system (CNS) they must: (a) enter the host, (b) spread from the site of entry to the CNS, (c) infect particular cells within nervous tissue, (d) cause death of infected neuronal cells, and (e) successfully avoid the host's immune defenses. Each of these aspects of CNS infection is discussed in the individual sections of this chapter. Additional information concerning the entry of reoviruses into the host, reovirus-induced apoptosis, and immunological responses to reovirus infection are covered in other chapters of this volume.

2 Entry

The name reovirus was originally proposed by SABIN (1959) to reflect the fact that viruses of this group had been isolated from the respiratory and enteric tracts, and were orphan ("reo") viruses without known associated disease. It is presumed from these early isolation studies that the respiratory and enteric tracts represent the natural portals of entry for reoviruses into the host.

2.1 Entry from the Gastrointestinal Tract

Using peroral inoculation of newborn mice as an experimental model system, the entry of reoviruses via the intestinal tract has been extensively investigated (WOLF et al. 1981, 1983, 1987; Bass et al. 1988; for review see chapter by Organ and Rubin, this volume). Virus is initially found associated with the luminal surface of specialized epithelial cells (M or microfold cells) that overlie submucosal collections of small intestinal lymphoid tissue (Peyer's patches) that form part of the system of gut-associated lymphoid tissue. Reoviruses appear to adhere specifically to the apical membranes of M cells (Amerongen et al. 1994). Virion particles can be sequentially followed as they bind to the luminal surface of M cells, and are then transported across these cells within endocytic vesicles to their basolateral surface (Wolf et al. 1981, 1983, 1987). The binding of reoviruses to M cells is materially enhanced by the intraluminal digestion of virion particles into intermediate subviral particles through the action of intestinal proteolytic enzymes (Bass et al. 1990; Amerongen et al. 1994). It has been proposed that M cells are an integral part of the intestinal immune system whose normal function is to facilitate transfer and subsequent presentation of ingested antigens to intestinal lymphoid cells (Neutra and Kraehenbuhl 1992). Although reoviruses were one of the first pathogens found to enter the host by penetrating the intestinal epithelial layer using the M-cell pathway, it has subsequently been recognized that M cells also serve as the site of entry for polioviruses (Sicinski et al. 1990), human immunodeficiency virus (Amerongen et al. 1991), and a variety of enteric bacteria (Morrison and Fields 1991).

2.2 Entry from the Respiratory Tract

Entry of reoviruses via the respiratory tract appears to follow a pathway analogous to that seen in the intestine (Morin et al. 1994, 1996). M cells are also found in the respiratory epithelium overlying patches of bronchus-associated lymphoid tissue. Reovirus serotype 1 Lang (T1L) inoculated intratracheally into adult rats binds to the luminal surface of pulmonary M cells and is transported transepithelially to reach the basal surface of these cells. Within 30 min postinoculation virus can be detected intracellularly in membrane-bound vesicles, and by 1-h virion particles are beginning to appear on the basal surface beneath the respiratory epithelium.

2.3 Other Routes of Entry

Experimental models of reovirus infection frequently exploit "nonnatural" routes of viral entry to initiate CNS infection. These routes of entry include direct intracranial, footpad or intramuscular, oronasal, and intraperitoneal inoculation. Following footpad inoculation of reovirus T1L or serotype 3 Dearing (T3D) virus can be detected in the epidermis and muscle. T1L is also found in the endothelium of blood vessels (Flamand et al. 1991). Both subcutaneous and intraperitoneal inoculation of suckling hamsters or ferrets with T1L results in the efficient spread of virus to the brain and spinal cord and the subsequent generation of ependymitis and hydrocephalus (see below; Kilham and Margolis 1969; Margolis and Kilham 1969a,b; Nielsen and Baringer 1972; Milhorat and Kotzen 1994).

Intraperitoneal inoculation of T3 in immunocompetent neonatal mice also allows for viral dissemination to a variety of organs including brain and the subsequent development of encephalitis (STANLEY et al. 1953; PAPADIMITRIOU 1967; SPRIGGS et al. 1983). Similar results have been described in adult immunocompromised mice. Following intraperitoneal inoculation of adult CB17 SCID mice with either T1L or T3D virus can be detected in a variety of organs, including brain (HALLER et al. 1995a,b; see below). The exact cellular pathways of viral entry following intraperitoneal and oronasal inoculation of virus have not been established.

3 Spread

In order to produce neurological disease, reoviruses must spread from their original site of entry into the host to reach the CNS. The pathways of reovirus spread have been extensively investigated following inoculation of virus both by the footpad route (TYLER et al. 1986; FLAMAND et al. 1991) and following oral inoculation (KAUFFMAN et al. 1983; MORRISON et al. 1991).

3.1 Spread Following Subcutaneous or Intramuscular Inoculation

Following footpad inoculation of T1L into newborn mice virus appears with identical kinetics and similar titer in all regions of the spinal cord (Fig. 1, panels a, b). This spread is not inhibited either by denervation of the inoculated limb or by treatment of animals with pharmacological agents that disrupt axonal transport (Fig. 1 panel c, Fig. 2 panel c) (TYLER et al. 1986). Viral antigen is consistently detected in the endothelium of both arteries and veins (KUNDIN et al. 1966; FLAMAND et al. 1991). An early low titer viremia is followed by a progressive increase in amount of virus present in the bloodstream (Fig. 3) (FLAMAND et al. 1991), although this declines after the second week of illness (KUNDIN et al. 1966). Thus the principal mode of spread of T1L to the CNS is through the bloodstream (hematogenous spread). By contrast, following footpad inoculation of T3D virus is detected first and in significantly greater amounts in the region of the spinal cord containing the neurons innervating the site of inoculation (Fig. 4 panels a, b) (TYLER et al. 1986). Denervation of the inoculated limb inhibits spread of T3D, as does treatment of animals with colchicine, a pharmacological inhibitor of the microtubule-associated pathway of fast axonal transport (Fig. 2 panels a, b, Fig. 4, panel c) (TYLER et al. 1986). Viral antigen can be detected in motor neurons within the spinal cord as early as 14h postinfection and in sensory neurons within 19h postinfection. Antigen is initially detected in cells ipsilateral to the site of inoculation. Both the amount of virus and the number of T3D antigen-positive cells increases progressively at subsequent times postinfection (TYLER et al. 1986; FLAMAND et al. 1991). Thus the principal pathway of spread of T3D to the CNS

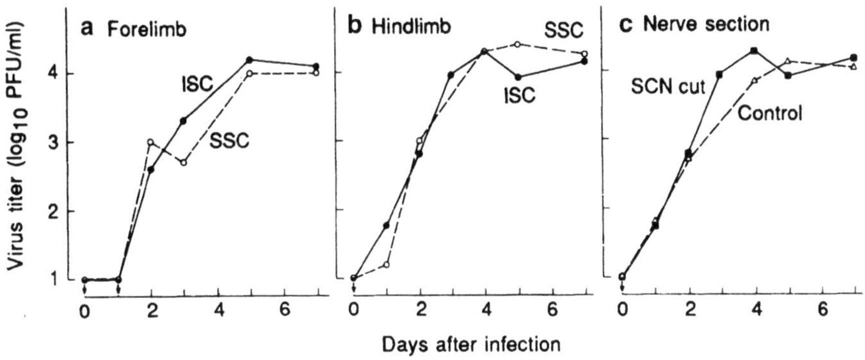

Fig. 1. Pattern of spread of reovirus type 1 to the spinal cord of neonatal mice after inoculation of the virus into the forelimb (*a*) or hindlimb (*b*) footpad. *Open circle*, Superior spinal cord (*SSC*); *closed circle*, inferior spinal cord (*ISC*). *c* Spread of type 1 to the ISC after sectioning of the sciatic nerve (*closed square*) is compared with spread in control animals with intact nerves (*open triangle*). *Downward-pointing arrow*, (*below data point*) no virus was detected at the lowest dilution. (From TYLER et al. 1986)

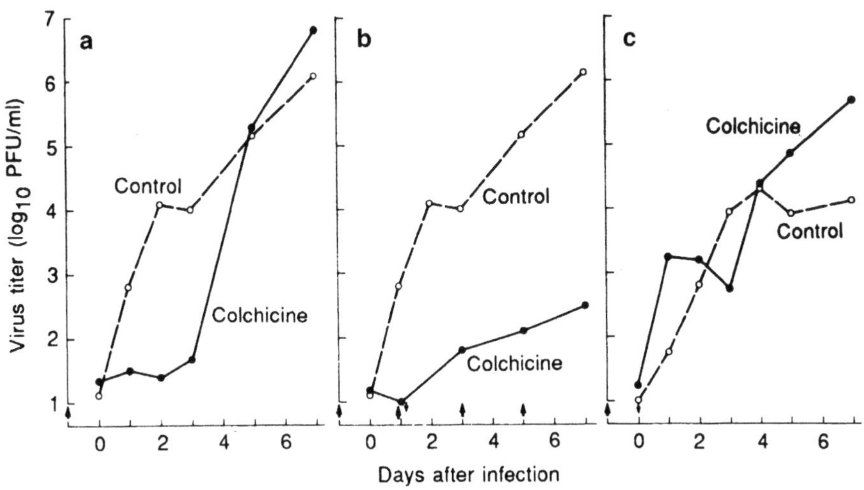

Fig. 2. Pattern of spread of type 3 (*a* and *b*) or type 1 (*c*) to the inferior spinal cord in colchine-treated (*closed circle*) and untreated control (*open circle*) mice. *Upward-pointing arrows* (*above time scale*), time of colchine administration. (From TYLER et al. 1986)

following peripheral inoculation is by neural spread and the utilization of the intraneuronal system of fast axonal transport.

The neural spread of T3D can involve either retrograde transport within the axons of motor neurons or anterograde transport within the axons of sensory neurons (Fig. 5). For example, following subcutaneous inoculation of reovirus serotype 3 clone 9 (T3C9) into the forehead of newborn mice, virus is detected within the motor neurons of the VIIth (facial) cranial nerve nucleus and within sensory neurons of the Vth (trigeminal) nucleus (MORRISON et al. 1991).

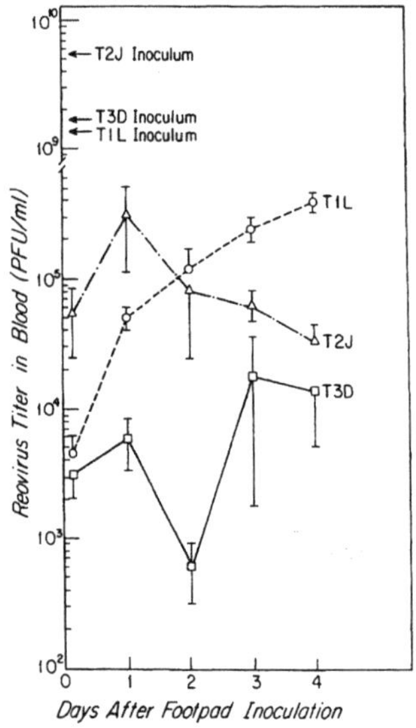

Fig. 3. Virus in the blood after footpad inoculation with T1L, T2J, or T3D. Two-day-old mice were inoculated in the right footpad with 5μl purified virus and were killed 3, 24, 48, 72, or 96h after inoculation. Blood samples of 50μl were collected, and the titer of virus in the blood and in the inocula (*arrows*) was determined by plaque assay. *Plotted points*, mean titer of virus from four animals with corresponding standard errors corrected for the small sample size. *Open circle*, T1L; *open triangle*, T2J; *open square*, T3D. (From FLAMAND et al. 1991)

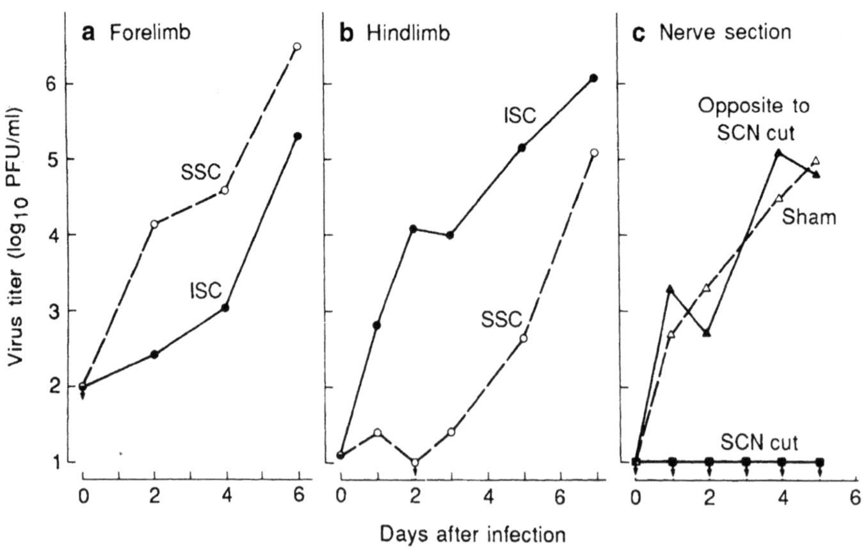

Fig. 4. Pattern of spread of reovirus type 3 to the spinal cord of neonatal mice after inoculation of the virus into either the forelimb (*a*) or hindlimb (*b*) footpad. *Open circle*, Superior spinal cord (*SSC*); *closed circle*, inferior spinal cord (*ISC*). *c* Sciatic nerve section in limb ipsalateral (*closed square*) or contralateral (*closed triangle*) to virus inoculation; *open triangle*, sham operation. *Downward-pointing arrow*, no virus was detected at the lowest dilution shown. (From TYLER et al. 1986)

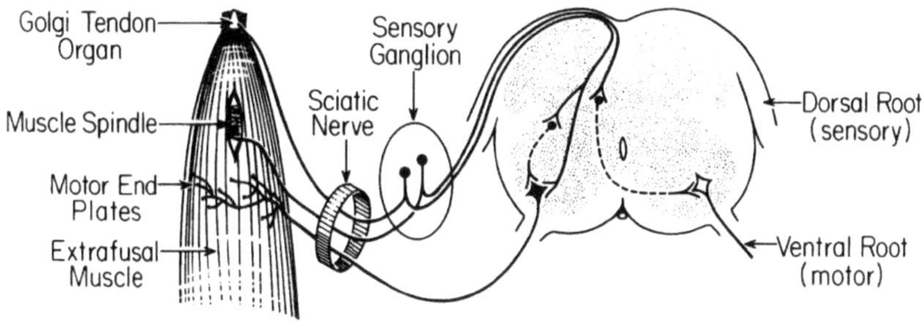

Fig. 5. Schematic representation of potential sensory and motor neuronal pathways traveled by reovirus from hindlimb muscles to the lumbar spinal cord. Motor innervation of the muscle spindles as well as most connections of sensory and motor neurons have been omitted for clarity. *Solid lines*, sensory and motor nerve fibers; *dashed lines*, interneurons; *filled cell bodies*, categories of neurons which could be primarily or secondarily infected. (From FLAMAND et al. 1991)

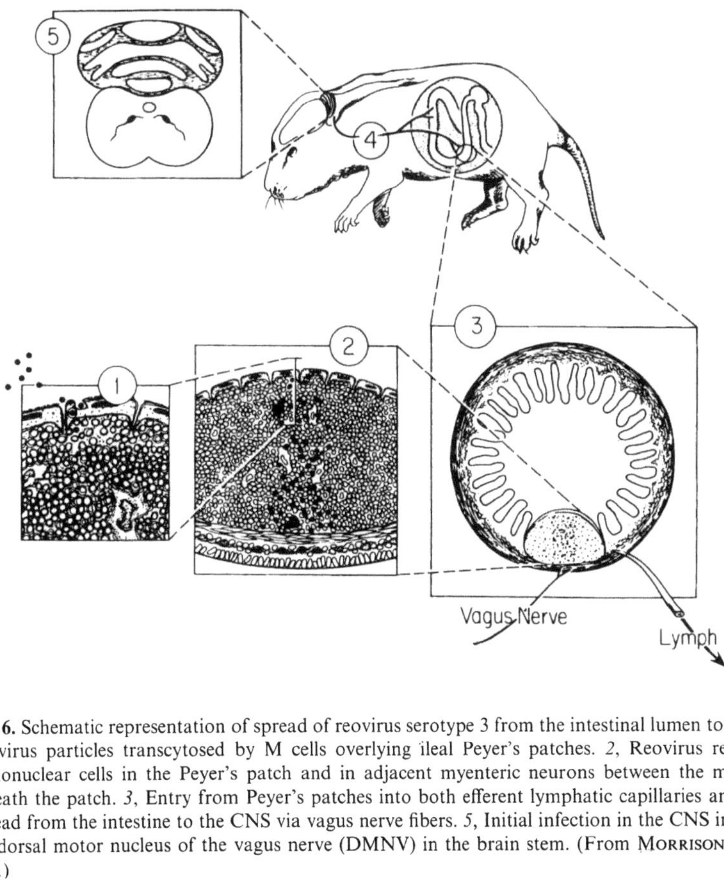

Fig. 6. Schematic representation of spread of reovirus serotype 3 from the intestinal lumen to the CNS. *1*, Reovirus particles transcytosed by M cells overlying ileal Peyer's patches. *2*, Reovirus replication in mononuclear cells in the Peyer's patch and in adjacent myenteric neurons between the muscle layers beneath the patch. *3*, Entry from Peyer's patches into both efferent lymphatic capillaries and nerves. *4*, Spread from the intestine to the CNS via vagus nerve fibers. *5*, Initial infection in the CNS in neurons of the dorsal motor nucleus of the vagus nerve (DMNV) in the brain stem. (From MORRISON and FIELDS 1991)

3.2 Spread from the Gastrointestinal Tract

The pathway of spread of reoviruses from the intestinal lumen has been investigated by immunohistochemistry (Figs. 6, 7) (MORRISON et al. 1991). This route of spread is of particular significance as infection by the oral route represents a natural portal of entry for reoviruses. The small intestine is extensively innervated by autonomic nerve fibers. Preganglionic parasympathetic efferent nerve fibers arise from neurons in the dorsal motor nucleus of the vagus nerve within the medulla of the brainstem. These fibers travel within the vagus nerve and end in plexuses containing ganglion cells within or near the organs being innervated. In the case of the small intestine these fibers terminate in neurons within the myenteric plexus, which is located between the circular and longitudinal muscular layers of the intestine. Preganglionic sympathetic efferent fibers to the small intestine originate in neurons within the intermediolateral cell column of the thoracic and lumbar spinal cord. These fibers leave the spinal cord via the ventral roots and enter the sympathetic trunk and the greater splanchnic nerve. These fibers ultimately synapse on postganglionic neurons in the celiac and mesenteric ganglia.

After purified reovirus T3C9 is orally inoculated into 2-day-old NIH Swiss mice, viral antigen is detected within 3 days postinfection throughout mononuclear cells in the lymphoid follicles of Peyer's patches within the terminal ileum (MOR-

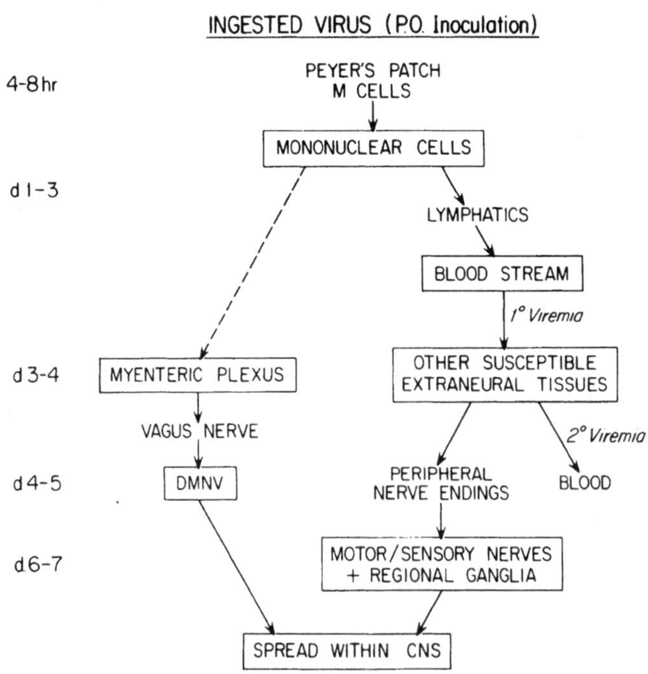

Fig. 7. Schematic representation of steps in reovirus T3C9 pathogenesis. *Left*, temporal progression. (From MORRISON et al. 1991)

RISON et al. 1991). Antigen is also detected in neurons of the myenteric plexus of the intestine beneath or adjacent to the margins of these infected Peyer's patches. Four days postinfection viral antigen is found in neurons of the dorsal motor nucleus of the vagus nerve in the medulla, but not in the spinal cord. This pattern of spread indicates that T3C9 spreads from the intestine using neural pathways that initially involves infection of the parasympathetic neurons in the myenteric plexus, virus then enters the axons of preganglionic parasympathetic neurons and spreads via these axons, which are contained within the vagus nerve, to reach neuronal cell bodies in the brainstem. These studies indicate that, in addition to spreading via motor and sensory neurons, virus can also be transported within neurons of the autonomic nervous system.

3.3 Genes Determining Patterns of Spread

Studies using reassortant viruses containing different combinations of genes derived from T1L and T3D indicate that the viral S1 gene, which encodes the structural protein σ1 and a smaller nonvirion associated protein (σ1s), is the primary determinant of the predominant pathway of spread of these viruses to the CNS following footpad inoculation (TYLER et al. 1986). The same gene also determines differences in the capacity of purified reovirions of different strains to associate with isolated preparations of microtubules derived from the brains of chicks or rabbits, or the cytoplasm of HeLa cells (BABISS et al. 1979). This is an interesting observation, given the repeatedly recognized propensity of reovirions to associate with microtubules in infected cells and the key role played by the cytoplasmic microtubular system as the scaffolding for fast axonal transport within neurons. The parallel between the in vivo studies of neural spread and the in vitro studies on microtubules is, however, imperfect. In vivo it is the T3D-derived S1 gene which determines the propensity of viruses to use neural spread, whereas in vitro it appears to be the T1L-derived S1 gene which determines enhanced efficiency of microtubule association. Nonetheless these findings do suggest that the viral σ1 protein is important in virion-associated microtubule association, a property which may find its ultimate expression in the neural spread of T3D.

The S1 gene also determines the pattern of viral spread from the gastrointestinal tract (KAUFFMAN et al. 1983), suggesting that it may have a more global role in determining the pattern of spread of reoviruses in the infected host. Following peroral inoculation of 6- to 8-week-old C3H/HeJ or 10-day-old BALB/cJ mice with T1L, virus can be sequentially found in Peyer's patches and small bowel followed by mesenteric lymph nodes and finally the spleen. This pattern of spread suggests that T1L initially entered local lymphatics with subsequent bloodstream invasion. By contrast, T3D, although also initially entering Peyer's patches, is not detectable in mesenteric lymph nodes or spleen. Using reassortant viruses containing different combinations of genes derived from T1L and T3D, it is possible to show that the capacity of these viruses to spread to extraintestinal organs is determined by the S1 gene.

In an attempt to further examine the role of the S1 gene in determining patterns of spread, reovirus variants containing single amino acid substitutions in the T3D $\sigma 1$ protein have been studied (see below; SPRIGGS and FIELDS 1982; SPRIGGS et al. 1983; BASSEL-DUBY et al. 1986; KAYE et al. 1986). Following intraperitoneal inoculation these variants show similar patterns of growth in liver and spleen but fail consistently to spread to or grow within the CNS (SPRIGGS et al. 1983). Following peroral inoculation the variant virus "K" shows a similar pattern of growth in intestinal tissue and subsequent spread to spleen as wild-type T3D, but, unlike T3D, variant K fails to reach the CNS (KAYE et al. 1986). Following subcutaneous or intracerebral inoculation the variant K shows the same pattern of neural spread as wild-type T3D but has an attenuated pattern of growth in both the spinal cord and retina (KAYE et al. 1986), similar to that seen in the brain (SPRIGGS and FIELDS 1982). These results suggest that mutations in the globular head of the $\sigma 1$ protein of T3D also restrict the spread of these viruses, but that when spread does occur, it follows the basic pattern seen with wild-type T3D

3.4 Neural and Hematogenous Spread

It is important to recognize that the use of neural and hematogenous spread need not be mutually exclusive processes. Immunocytochemical studies indicate that following footpad inoculation T1L antigen can be detected in the cell bodies of a few motor and sensory neurons within the spinal cord, although it remains unclear whether virus reaches these cells after initiating infection at their periphery or after reaching their cell bodies in the spinal cord through the bloodstream (FLAMAND et al. 1991). T3D is also capable of generating a viremia after footpad inoculation, but this does not increase significantly over time and remains nearly 100-fold lower than the amount of viremia generated by an equivalent dose of T1L (Fig. 3) (FLAMAND et al. 1991). These results suggest that the facility with which reoviruses spread to the CNS should perhaps be considered along a continuum, with some strains being more dependent on viremia and less on neural spread (T1L) and others more dependent on neural spread and less on viremia (T3D).

3.5 Spread Following Intraperitoneal and Oronasal Inoculation

Reoviruses are also capable of reaching the CNS following both oronasal (BARTHOLD et al. 1993) and intraperitoneal inoculation (STANLEY et al. 1953, 1964; WALTERS et al. 1963, 1965; PAPADIMITRIOU 1967; DAVIS 1982; SPRIGGS et al. 1983; HALLER et al. 1995a,b). Oronasal inoculation of T1 strains has been used as an experimental model for inducing ependymitis and hydrocephalus (KILHAM and MARGOLIS 1969; MARGOLIS and KILHAM 1969a,b; NIELSEN and BARINGER 1972; see below). Following oronasal inoculation T1L typically appears in brain between 2 and 5 days postinoculation, with peak viral titers around day 7. Virus titers decline from this point, and 14–21 days virus is no longer detectable in mouse brain (BARTHOLD et al. 1993).

Reovirus T3 also spreads to the brain following oronasal inoculation, although this route of spread has been less extensively studied than with T1. Peak viral titers generally occur around 9 days postinoculation. Mice are obviously ill by 12 days postinoculation and show histological evidence of a severe panencephalitis (BARTHOLD et al. 1993; see below). Unfortunately, the exact route of spread used by T3 reoviruses to reach the CNS after inoculation by this route remains unknown. In particular, the potential contribution of spread via the olfactory or trigeminal nerves has not been established.

Reoviruses can also reach the CNS after intraperitoneal inoculation in immunocompetent newborn mice (STANLEY et al. 1953, 1954; WALTERS et al. 1965; PAPADIMITRIOU 1967; KILHAM and MARGOLIS 1969; SPRIGGS et al. 1983), hamsters (DAVIS 1982), and immunocompromised adult CB17 SCID mice (HALLER et al. 1995a,b). Following intraperitoneal inoculation of T3D peak T3 viral titers in newborn mice occur around day 9 postinfection and are typically 10^8PFU/brain (SPRIGGS et al. 1983). Mice may show clinical signs of illness as early as 5–6 days postinoculation (STANLEY et al. 1953, 1954). Following intraperitoneal inoculation of T1L peak viral titers in neonatal hamsters, rats, and mice generally occur 2–3 days postinoculation, and in neonatal ferrets 4–5 days postinoculation (KILHAM and MARGOLIS 1969). In SCID mice intraperitoneal inoculation of T1L results in substantially higher brain titers than inoculation of T3D. Brain titers 12 days postinoculation are approximately $10^{6.5}$PFU/brain for T1L and 10^3PFU/brain for T3D (HALLER et al. 1995a,b). T3-infected mice may survive for prolonged periods, with an associated steady increase in viral titer in brain. Titers in mice surviving for 3 months postinoculation reach $10^{7.5}$PFU/brain (HALLER et al. 1995b).

The route of spread utilized by reoviruses to reach the CNS after intraperitoneal inoculation has not been established. However, in one early study T3 hepato-encephalomyelitis virus (HEV) was found in both the endothelium of cerebral capillaries and inside lymphocytes within the lumens of these capillaries, raising the possibility that this T3 strain reached the brain via the bloodstream after intraperitoneal inoculation (PAPADIMITRIOU 1967). A similar route of spread was postulated in studies of inner ear infection induced by serotype 3 clone Abney (T3A) in neonatal hamsters following intraperitoneal inoculation (DAVIS 1982).

3.6 Host Factors Influencing Spread

The age of animals at the time of reovirus infection influences spread of reoviruses from the periphery to the CNS (TARDIEU et al. 1983). Following subcutaneous inoculation of mice younger than 5 days of age with 5×10^8 of T3D, virus spreads to the brain (TARDIEU et al. 1983). Similar subcutaneous or intravenous inoculations in adult mice fail to produce detectable virus in the brain (TARDIEU et al. 1983). Whether these differences reflect differences in peripheral replication levels, patterns or mechanisms of spread, or susceptibility of neuronal tissue to infection are not known.

4 Tropism

Once reoviruses reach the CNS, the pattern of injury produced depends on the specific populations of cells infected.

4.1 The Reovirus Receptor

The reovirus receptor is discussed elsewhere (see chapter by SARAGOVI et al., vol. 1). Interest in this subject from the neuroscience point of view was stimulated by reports that there are structural similarities between the mammalian β-adrenergic and reovirus T3 receptors (Co et al. 1985a,b). This observation was based on: (a) the capacity of certain anti-reovirus receptor antibodies to immunoprecipitate purified β-adrenergic receptor, (b) the similarity in molecular mass and isoelectric points of the two receptors, (c) the similarity in the tryptic digest patterns of the two receptors, and (d) the fact that the T3 receptor is able to bind a β-receptor antagonist (iodohydroxybenzylpindolol), and that this binding is blocked by the β-receptor antagonist isoproterenol (Co et al. 1985b). Subsequent studies clearly demonstrate that T3 can infect cells that lack β-adrenergic receptor activity (SAWUTZ et al. 1987), and that T3 binding to cells with β-adrenergic receptor activity does not result in stimulation of cellular adenylate cyclase activity or the consequent increases in levels of cellular cAMP (SAWUTZ et al. 1987; CHOI and LEE 1988; DONTA and SHANLEY 1990). It has also been shown that binding of T3 does not interfere with the response of cells to β-adrenergic agonists (CHOI and LEE 1988; DONTA and SHANLEY 1990), nor does it result in the sequestration (down-regulation) of β-adrenergic receptors from the cell surface (CHOI and LEE 1988). This led to the suggestion that T3 and β-receptor ligands bind to distinct domains on the β-adrenergic receptor (LIU et al. 1988). This binding might involve areas near β-adrenergic receptor antagonist binding domains (DONTA and SHANLEY 1990). Following these studies the subject of the role of β-adrenergic receptors as putative reovirus receptors has received little recent attention or further study, and the issue must, as such, be considered as a still incompletely resolved area.

More recent investigations of the reovirus receptor have focused on potential roles for as yet unidentified cell surface proteins (ARMSTRONG et al. 1984; GENTSCH and HATFIELD 1984; CHOI et al. 1990) and the epidermal growth factor receptor as reovirus receptors (STRONG et al. 1993; TANG et al. 1993). There has also been renewed interest in the importance of sialic acid residues as binding sites for virus, and in identification of the regions of the σ1 cell attachment protein involved in interactions with these residues (GENTSCH and PACITTI 1985, 1987; PAUL and LEE 1987; PAUL et al. 1989; DERMODY et al. 1990; RUBIN et al. 1992; CHAPPELL et al. 1997). The relevance of these observations for the nature of the reovirus receptor on neuronal cells remains uncertain, and until further studies are conducted, the nature of the reovirus receptor on neuronal cells should be considered unknown.

4.2 CNS Infection by T1

Reoviruses belonging to serotype 1 (T1) and serotype 3 (T3) produce distinct patterns of infection within the CNS. Following intracerebral, intraperitoneal, subcutaneous, or oronasal inoculation into newborn mice, hamsters, rats, or ferrets (WALTERS et al. 1965; KILHAM and MARGOLIS 1969; MARGOLIS and KILHAM 1969a,b) T1L infects the ependymal cells lining the central canal of the spinal cord (MILHORAT and KOTZEN 1994) and the cerebral aqueducts and ventricles to produce ventricular enlargement (hydrocephalus) (Figs. 8, 9) (KILHAM and MARGOLIS 1969; MARGOLIS and KILHAM 1969a,b; PHILLIPS et al. 1970; NIELSEN and BARINGER 1972; MILHORAT and KOTZEN 1994). The incidence of hydrocephalus varies with the dose of virus, route of inoculation, and viral strain. Direct intracerebral inoculation is more effective than oronasal instillation in inducing hydrocephalus, and the incidence of hydrocephalus increases with increasing dose of virus (MASTERS et al. 1977).

4.2.1 Differences Between T1 Strains

Some variations in the temporal profile and severity of disease may occur between different T1 strains, although studies in this area are fragmentary at best. KILHAM and MARGOLIS (KILHAM and MARGOLIS 1969; MARGOLIS and KILHAM 1969a) compared the capacity of reovirus T1 strains T1L, SV12, and CVA to induce

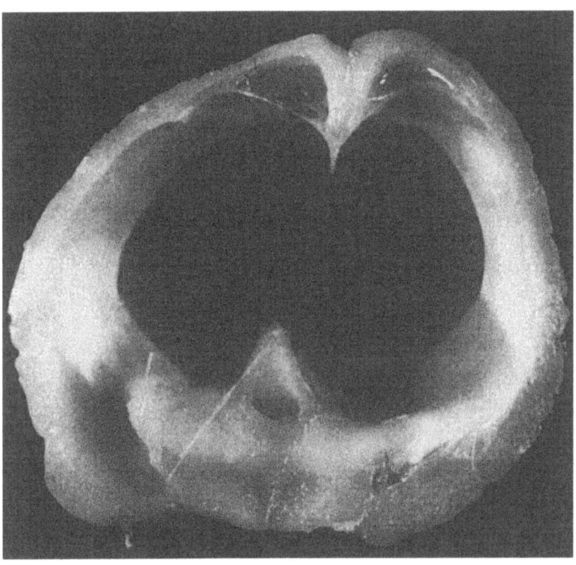

Fig. 8. Gross appearance of hydrocephalus 28 days after intraperitoneal infection of a neonatal hamster with T1L. (From NIELSEN and BARINGER 1972)

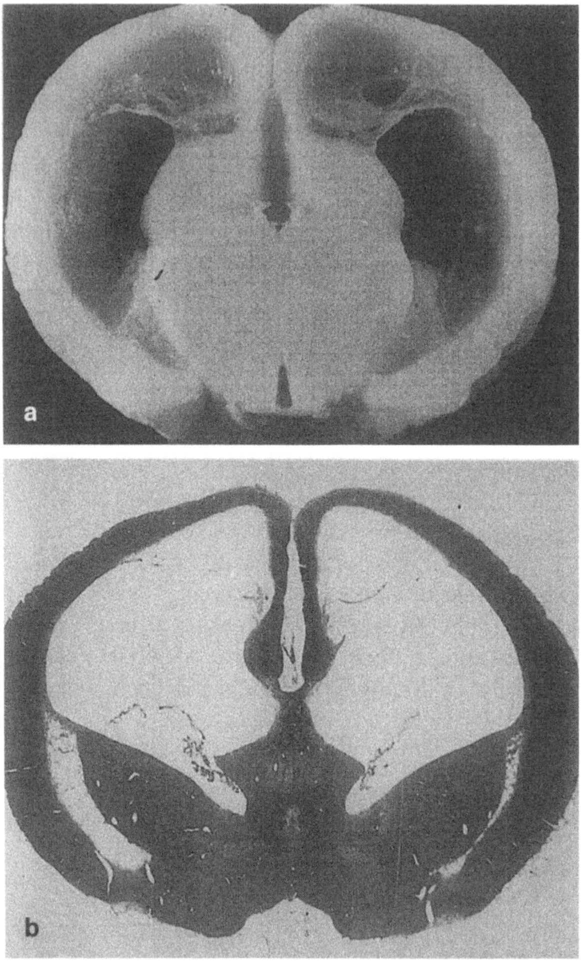

Fig. 9a,b. Coronal sections of brain. 3–4 weeks after intracerebral inoculation of suckling hamsters with T1L. **a** Moderately advanced hydrocephalus involving lateral and third ventricles. **b** Severe hydrocephalus. Hematoxylin-eosin, ×6. (From MARGOLIS and KILHAM 1969a)

ependymitis and produce hydrocephalus. Both T1L and CVA seemed to be more potent inducers of active ependymitis than SV12, although the time points used for comparison were not identical. However, SV12 appeared to induce more severe hydrocephalus than either T1L or CVA, although, again, matched temporal comparisons were not carried out. Overall the incidence of hydrocephalus varied from nearly 100% in hamsters inoculated with T1L to 56% in those inoculated with T1CVA (MARGOLIS and KILHAM 1969a). In another study a reovirus T1 strain obtained from SABIN (T1-S) produced less severe hydrocephalus than a T1 strain obtained from KILHAM (T1-K; MASTERS et al. 1977).

4.2.2 Clinical Features of T1 Disease

The clinical manifestations of reovirus T1 infection are somewhat variable. Mice surviving the acute infection can develop notable enlargement of the head (macrocephaly) secondary to hydrocephalus, although this is not a sensitive indicator of disease (PHILLIPS et al. 1970; MASTERS et al. 1977). Early neurological signs of infection can include incoordination and impaired walking, which appears to be a combination of both ataxia and impaired ability to properly use the hindlimbs. The ataxia and incoordination may reflect injury to the cerebellum resulting from its compression and downward displacement secondary to hydrocephalus (see below). Impaired hindlimb use may reflect stretching of corticospinal track fibers by the ventricular enlargement.

4.2.3 T1 Viral Titers and Kinetics of Growth

Following intracerebral inoculation of T1L viral titers generally peak between days 5 and 9 postinfection and then decline. Peak viral titers do not typically exceed 5×10^7PFU/brain with T1L (WEINER et al. 1980; HRDY et al. 1982; TARDIEU et al. 1983) but may reach 10^{8-9}PFU/brain with some T1L × T3D reassortant viruses capable of producing hydrocephalus (WEINER et al. 1980; see below). Most mice survive infection with T1L, and the LD_{50} following intracerebral inoculation has been reported to exceed 10^5PFU/mouse (WEINER et al. 1977), 10^6PFU/mouse (HRDY et al. 1982), and probably approximates 10^7PFU/mouse (TYLER et al. 1993). Footpad and oronasal inoculations of T1L generally do not produce lethal disease (TYLER et al. 1986; BARTHOLD et al. 1993).

4.2.4 T1-Induced Ependymitis and Hydrocephalus

T1 strains can infect both ependymal and choroid plexus epithelial cells (MARGOLIS and KILHAM 1969; MILHORAT and KOTZEN 1994). Viral antigen can be detected within these cells and can take the form of viral inclusions (KUNDIN et al. 1966; MARGOLIS and KILHAM 1969a,b; NIELSEN and BARINGER 1972; MILHORAT and KOTZEN 1994). Large aggregates of virions can be found in the cytoplasm of ependymal cells by electron microscopy. Dead ependymal cells are frequently sloughed into the lumen of the ventricles and aqueducts, leaving widespread areas of denuded ventricular surface (MARGOLIS and KILHAM 1969a,b; NIELSEN and BARINGER 1972; MASTERS et al. 1977) (acute ependymitis).

There appear to be several mechanisms leading to the development of hydrocephalus in T1-infected animals. Sloughed ependymal cells can mechanically obstruct the aqueducts and foramina connecting the ventricles. Regeneration of ependymal cells can occur but is often disorganized, resulting in discontinuous ependymal lining of the ventricles or buried clusters of cells (MARGOLIS and KILHAM 1969a,b). Obstruction to CSF circulation can also occur as a consequence of glial and vascular proliferation at sites of ependymal injury (MARGOLIS and KILHAM 1969a,b; NIELSEN and BARINGER 1972; MILHORAT and KOTZEN 1994). Glial and capillary proliferation tends to be maximal at sites of ependymal ulceration and can be quite pronounced. Proliferation of astroglia and capillary endothelial cells results in nodules, sheets, or bridging fibrous bands of tissue. These bands can obstruct CSF circulation through the foramina of Monro, the IIIrd ventricle, the aqueduct of Sylvius, the foramina of Luschka and Magendie, and the central canal of the spinal cord (MARGOLIS and KILHAM 1969a,b). Obstruction of the aqueduct of Sylvius can result in noncommunicating hydrocephalus with massive enlargement of the lateral and IIIrd ventricles and a relatively normal appearing IVth ventricle. Finally, induction of an intense basilar and cisternal meningitis can block CSF outflow from the IVth ventricle with production of communicating hydrocephalus and enlargement of the entire ventricular system. Inflammation can involve the arachnoid villi leading to arachnoiditis and ultimately arachnoid villus fibrosis (MASTERS et al. 1977). This results in impaired CSF readsorption at the arachnoid villi within dural venous sinuses. Each of these causes of hydrocephalus can be associated with thinning of the cortical mantle and with rostrocaudal displacement and compression of posterior fossa structures, including the brainstem and cerebellum (MARGOLIS and KILHAM 1969a,b; MASTERS et al. 1977).

It should be emphasized that it may take some time for the pathological process leading to hydrocephalus to take its course. As a result hydrocephalus may not be present acutely but may take several weeks or even months to develop fully. For example, following oronasal inoculation of T1L the time of onset of hydrocephalus ranged between 47–242 days (mean 109 days) (PHILLIPS et al. 1970). Intraperitoneal inoculation of neonatal hamsters with T1L resulted in hydrocephalus in all animals after 3 weeks (KILHAM and MARGOLIS 1969a,b; NIELSEN and BARINGER 1972). Following intracerebral inoculation of 10^7PFU of T1L hydrocephalus is found in more than 70% of mice surviving acute infection examined 28 days postinfection (TYLER et al. 1993). Others have noted a similar pattern of delay, with oronasal inoculation resulting in later development of hydrocephalus than intracranial inoculation (MASTERS et al. 1977).

A mononuclear inflammatory reaction involving both the ventricular CSF and the leptomeninges commonly occurs in T1-infected animals (MARGOLIS and KILHAM 1969a,b). Perivascular inflammation in the brain parenchyma is largely limited to the immediate periventricular areas (MARGOLIS and KILHAM 1969a,b). The peak of this inflammatory reaction generally occurs 10–12 days postinfection and then gradually subsides (MARGOLIS and KILHAM 1969a,b).

4.2.5 Neuronal and Pituitary Infection with T1

Neuronal infection is not a typical feature of infection with T1L, although isolated neurons are occasionally found to be infected (WALTERS et al. 1965; KUNDIN et al. 1966; MARGOLIS and KILHAM 1969a,b; MASTERS et al. 1977; WEINER et al. 1977; FLAMAND et al. 1991). Some early studies of infection with reovirus T1 strains, including T1 strain 716, describe more neuronal injury and infection, and provide less information concerning the presence of ependymitis and hydrocephalus (JENSON et al. 1965). Antigen-positive astrocytes and neurons have also been reported following subcutaneous infection with T1L (KUNDIN et al. 1966; FLAMAND et al. 1991). There are many differences between these older studies and more recent investigations, including variations in the viral strains, mouse species, routes of inoculation, and methods of observation employed. However, which if any of these factors accounts for the observed differences in ependymal versus neuronal tropism of T1 strains is unclear.

Following intracerebral inoculation into neonatal mice T1L infects the cells of the anterior pituitary (ONODERA et al. 1981). Within 7 days of infection viral antigen can be detected in the anterior lobe of the pituitary gland but not in the intermediate or posterior lobes. Pathological changes include an inflammatory cell infiltrate of mononuclear cells and evidence of "coagulation necrosis." Electron microscopy shows T1L virus particles within the cytoplasm of growth-hormone containing cells. Plasma levels of growth hormone are significantly decreased in infected animals, and this may account in part for their observed growth retardation (runting syndrome).

4.3 CNS Infection by T3

In distinction to the pattern of CNS infection produced by T1 strains, T3 strains produce a acute, fatal encephalitis with striking neuronal involvement and minimal if any involvement of ependymal cells (Fig. 10) (GONATAS et al. 1971; MARGOLIS et al. 1971; RAINE and FIELDS 1973; WEINER et al. 1977). Inflammatory responses may be minimal early, even when extensive neuronal destruction is present (MARGOLIS et al. 1971). At later times postinfection both perivascular, parenchymal, and leptomeningeal inflammation is present (RAINE and FIELDS 1973). The degree of inflammation may be striking, especially in the perihippocampal area, of mice surviving infection either because of administration of immunotherapy or initial inoculation with attenuated T3D variant viruses (TYLER et al. 1989). Early studies suggested with the HEV strain of T3 suggested that virus could be found within cerebral capillary endothelial cells and in perivascular protoplasmic astrocytes (PAPADIMITRIOU 1967). This virus could also be seen within the cytoplasm, axons, and dendrites of neurons. Subsequent studies with the Dearing strain of T3 suggested that this virus is not typically found in cerebral endothelial or perithelial cells, and that glial cell infection is minimal or absent (GONATAS et al. 1971; RAINE and FIELDS 1973). Ultrastructural studies revealed virus particles in neuronal perikarya and dendrites, but T3D virions were not specifically visualized in axons or presynaptic nerve terminals (GONATAS et al. 1971). Both T3D and T3HEV virions were seen in lysosomes within the cytoplasm of infected neurons.

4.3.1 T3 Viral Titers, Kinetics of Growth, and Lethality

Mice infected intracerebrally with T3 strains almost invariably die of lethal encephalitis (HRDY et al. 1982). The intracerebral LD_{50} in newborn mice for T3 strains varies between levels below 10^1 and 10^4PFU/mouse (HRDY et al. 1982). Death, following high-dose inoculums, can occur as early as 6 days postinfection,

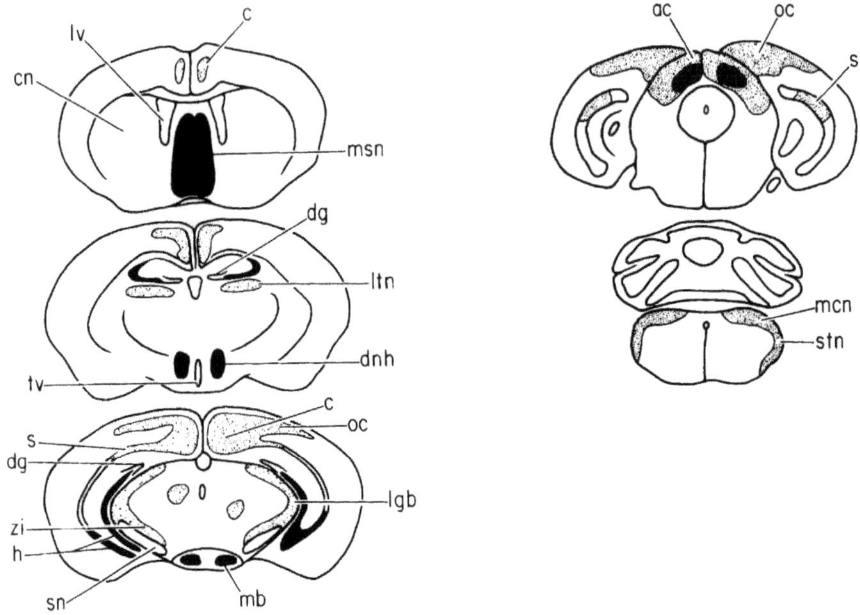

Fig. 10. Diagrams of coronal brain sections extending from rostral (*top left*) to caudal (*bottom right*) areas of the brain showing the location of virus-induced lesions in mice infected intracerebrally with either T3D or T3D variants with amino acid substitutions in the σ_1 cell attachment protein. *Black areas*, regions of the brain in which T3D virus as well as the variant viruses induce lesions; *dotted areas*, regions in which only the T3D virus cause necrosis. *ac*, Anterior colliculus; *c*, cingulum; *cn*, caudate nucleus; *dg*, dentate gyrus; *dnh*, dorsomedial nucleus of the hypothalamus; *h*, hippocampus; *lgb*, lateral geniculate body; *ltn*, lateral thalamic nucleus; *lv*, lateral ventricle; *mb*, mammillary bodies; *msn*, medial septus nucleus; *oc*, occipital cortex; *s*, subiculum; *sn*, substantia nigra; *stn*, spinal trigeminal nucleus; *tv*, third ventricle; *zi*, zona incerta. (From SPRIGGS et al. 1993)

and the majority of mice are dead within 10–11 days postinfection. Higher intracerebral inocula typically result in earlier mortality, with the mean day of death between days 6 and 7 at doses of higher than 10^5PFU/mouse, between days 9 and 10 at doses of 10^{2-3}PFU/mouse, and 11 days after 10PFU/mouse (WEINER et al. 1977; VIRGIN et al. 1988). Peak viral titers can exceed 10^{9-10}PFU/brain (WEINER et al. 1980; HRDY et al. 1982; VIRGIN et al. 1988; TYLER et al. 1989). Mice dying of T3D infection generally have brain titers exceeding 10^8PFU/brain (VIRGIN et al. 1988). The LD_{50} of T3D is considerably higher ($10^{3.7-4.8}$PFU/mouse) following footpad or intramuscular inoculation than after intracerebral inoculation (VIRGIN et al. 1988; HALLER et al. 1995), and the mean day of death is considerably later (10–15 days) (VIRGIN et al. 1988), although peak viral titers in brain are similar (TYLER et al. 1989). Although detailed studies have not been performed, when lethal doses of either T3D or T3A virus are inoculated, the severity and topographic distribution of lesions within the CNS do not appear to differ significantly, whether viruses are inoculated by the subcutaneous or intracerebral route (MARGOLIS et al. 1971; TYLER, unpublished).

4.3.2 T3 Encephalitis and Neuronal Infection

The typical pattern of infection with T3 viruses results in widespread injury within the CNS (Van Tongeren 1957; Walters et al. 1963; Stanley et al. 1964; Weiner et al. 1977; Spriggs et al. 1983; Tardieu et al. 1983; Tyler and Fields 1988, 1996; Tyler 1991; Morrison et al. 1993; Oberhaus et al. 1997; Virgin et al. 1997). As noted, predominantly neurons are affected; there is sparse involvement of astrocytes or microglia, sparing of oligodendrocytes, and typically only rare involvement of ependymal cells. Parenchymal changes are not initially accompanied by inflammatory changes (Morrison et al. 1993), although an inflammatory and astroglial reaction becomes more prominent in the later stages of disease and can be quite dramatic in mice surviving the acute neurological illness (Tyler et al. 1989). Within the cortex the involvement is typically patchy rather than diffuse, with a predilection for maximal involvement of cingulate gyrus and occipital lobes. Injury in the hippocampus is typically severe with maximal destruction of pyramidal neurons within CA2–CA4. Interestingly, injury may be less pronounced within CA1 and the dentate gyrus, although this is not invariable. Other areas of the limbic system including the subiculum, septal nuclei, and mammillary bodies are also severely injured. Thalamic nuclei are frequently involved, and to a lesser degree the basal ganglia. Of the thalamic nuclei, the lateral geniculate body, a major visual relay area, and the lateral thalamic nuclei are the most dramatically affected. Within the cerebellum the Purkinje cell layer often shows loss of cells, or cells containing prominent viral inclusions and staining positive for viral antigen. T3 also produces injury to the brainstem and spinal cord following intracerebral inoculation. Involvement is typically patchy, with affected areas including the superior colliculus in the midbrain and the cuneate nucleus and trigeminal nuclei within the medulla. Massive necrosis of retinal ganglion cells can occur, with striking inclusions (Tyler et al. 1985). Viral titers in the retina are among the highest encountered in the CNS (Tyler et al. 1985). Reovirus T3A has also been reported to infect the neurons of the cochlear and vestibular ganglia of the inner ear (Davis 1982).

4.3.3 T3-Induced Apoptosis in CNS In Vivo

Recent studies indicate that apoptosis may be an important mechanism of T3D-induced injury to neural tissue (Oberhaus et al. 1997; for review, see chapter by Oberhaus et al., this volume). Following intracerebral inoculation of T3D, oligonucleosomal laddering of cellular DNA extracted from brain can be detected. This appears to be maximal 8–9 days postinfection and corresponds to the period at which maximal brain titers of virus are achieved. Within the brain neuropathological injury, viral antigen distribution, and apoptosis colocalize to the same brain regions. The majority of apoptotic cells, detected by terminal deoxynucleotidyl transferase (TdT) mediated dUTP-biotin nick end-labeling (TUNEL) staining, appear to be neurons. Apoptotic cells are seen only in regions of viral infection. All regions of the brain in which neuropathological injury occurs contain both antigen-positive and TUNEL- positive cells. Within specific brain regions the relative proportion of apoptotic to infected cells varies, suggesting that cells within certain regions of the brain are more susceptible to apoptotic cell death. Apoptotic and infected cells occur in the cingulate gyrus and occipital cortex, the thalamus, and hippocampus. Examination of individual cells within these regions indicates that cells may be both infected (antigen positive) and apoptotic (TUNEL positive), apoptotic but not antigen positive, or antigen positive but not apoptotic. These results suggest that apoptosis may be either a direct process occurring in infected cells (antigen positive and TUNEL positive), or an indirect process (antigen negative, TUNEL positive) occurring in cells in the immediate vicinity of infected cells.

4.3.4 Clinical Features of T3 Disease

Clinical features of reovirus encephalitis are relatively stereotyped (Stanley et al. 1953; Stanley et al. 1954; Walters et al. 1963; Stanley et al. 1964). The progression of illness is typically rapid, with death occurring in the majority of animals within 48h of the onset of symptoms. In some cases mice die even within hours of the onset of symptoms (Stanley et al. 1954). Initial symptoms consist of enhanced tremulousness, shakiness, and incoordination (Walters et al. 1963; Stanley et al. 1964). Many animals develop focal or generalized seizures characterized by rhythmic limb jerking. Paralysis of one or more limbs may occur. In the late stages of infection mice are lethargic, often lying immobile in the cage, with the head anteflexed onto the chest, and the spine arched in flexion. Occasional mice show a brief period of hyperexcitability and may race around the cage if disturbed. Mice with encephalitis frequently appear small, runted, or emaciated compared to healthier littermates, perhaps due to impaired ability to maintain adequate nutrition and hydration and in some cases to the contributory coexistence of gastrointestinal and hepatobiliary disease.

4.4 CNS Infection by T2

The pattern of CNS infection produced by T2 strains and the pathogenesis of this infection have received little attention. Following peroral inoculation serotype 2 strain D5 (T2D5) produces injury to neurons in the thalamus and cerebellar peduncles (Walters et al. 1965). This is associated with a mononuclear cell infiltrate and perivascular cuffing. Infected mice usually survive, and clinical signs and symptoms are reportedly minimal or entirely absent (Walters et al. 1965). Following intraperitoneal inoculation T2D5 also reaches the CNS, producing a pattern of infection similar to that described following peroral inoculation. T2D5 has not been reported to induce ependymitis or hydrocephalus (Phillips et al. 1970), and as such its pattern of CNS infection appears to more closely resemble that of T3 rather than T1 reovirus strains (see above).

5 Prolonged Replication of Virus in the CNS

Under most circumstances infection in vivo with T3 reoviruses induces an acute, highly virulent and often lethal CNS infection. In cell culture reoviruses can produce persistent infections (see chapter by DERMODY, this volume), but there has never been a clear analogue to this phenomena described in vivo. Reoviruses which bear mutations associated with the establishment and maintenance of persistent infections (PI) in vitro have been tested for their capacity to induce persistent CNS infection in vivo (MORRISON et al. 1993). Sixteen independently derived T3D PI viruses were injected intracranially into 2-day-old NIH Swiss mice. Of the 16 PI isolates 12 had a lethality similarity to wild-type T3D. Three PI viruses were approximately threefold less lethal than wild-type T3D. The most attenuated of the PI viruses (PI 7–2) had an intracerebral LD_{50} approximately 56,000 times greater than T3D (MORRISON et al. 1993). The majority (84%) of mice surviving infection with wild-type T3D have no detectable virus in their brains by 25 days postinfection. Approximately 16% of these mice have residual titers that range from 10^2–10^5PFU/brain (mean titer $10^{3.3}$PFU/brain). By contrast, 38% of mice surviving infection with PI viruses have residual virus detectable in their brains, with the highest residual titers, in some cases reaching nearly 10^8PFU/brain (mean titer $10^{4.6}$) (MORRISON et al. 1993). Despite these apparent differences in titer, the anatomic distribution of viral antigen and of neuropathological injury does not differ significantly between the PI viruses and T3D (Fig. 11) (MORRISON et al. 1993). Thus, although viruses which are capable of producing PI in cell culture do show a pattern of prolonged replication within the mouse CNS following in vivo infection, they do not generate persistent infections in vivo.

6 CNS Infection in Immunocompromised Mice

Reovirus infection in immunocompromised mice provides another model for looking at neurotropism, neurovirulence, and the potential of reoviruses to generate prolonged infection within the CNS. Under normal circumstances (see above), mice demonstrate a clear age-dependent susceptibility to reovirus infection (TARDIEU et al. 1983). Following intracerebral inoculation of 5×10^5PFU T3D all mice inoculated at or before 8 days of age die (TARDIEU et al. 1983). Adult mice fail to succumb to intracerebral inoculation with 5×10^9PFU (TARDIEU et al. 1983) or even more massive (TYLER, unpublished) doses of T3D. Adult animals also fail to show histological evidence of CNS injury following intracerebral challenge with T3D (TARDIEU et al. 1983). Peak viral titers in brain of mice inoculated with either T1L or T3D decline progressively with increasing age of the inoculated animal. Animals up to 8 days of age show evidence of viral replication in the brain whereas mice older than 12 days do not (TARDIEU et al. 1983).

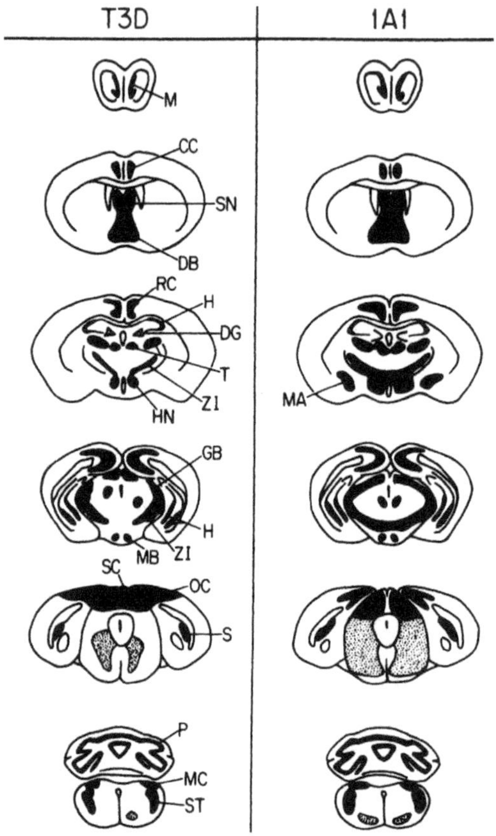

Fig. 11. Schematic representation of pathology in coronal sections of brain tissue obtained from mice 12 days after i.c. inoculation with a lethal dose of virus. *Left*, T3D-induced pathology; *right*, PI 1A1-induced pathology. *Solid shading*, areas of intense tissue destruction; *stippled shading*, areas of sparse pathological changes. *CC*, Cingulate cortex; *DB*, nucleus of the ventral limb of the diagonal band; *DG*, dentate gyrus, pyramidal and polymorpic layers; *GB*, lateral geniculate body; *H*, hippocampus pyramidal layer regions CA2 and CA3; *HN*, hypothalamic nuclei; *M*, mitral cell layer; *MA*, medial amygdaloid nucleus, anterior portion; *MB*, mamillary body; *MC*, medial cuneate nucleus; *OC*, occipital cortex; *P*, Purkinje cells; *RC*, retrosplenial cortex; *S*, subiculum; *SC*, superior colliculus; *SN*, septal nucleus, medial and lateral; *ST*, spinal trigeminal nucleus; *T*, thalamic nuclei, medial and lateral; *ZI*, zona inserta. (From Morrison et al. 1993)

By contrast to the results in immunocompetent adult mice, immunoincompetent adult mice can replicate virus, develop lethal infections, and show striking patterns of tissue injury (George et al. 1990; Haller et al. 1995a). When adult CB17 SCID mice are inoculated intraperitoneally with reovirus T1L, they die an average of 20 + 6 days (range 14–35 days) after infection. Mice infected with T3D die an average of 77 + 22 days (range 24–105 days) after infection. Virus titers in organs including brain, intestine, and liver are consistently higher in T1L than in T3D infected animals. Liver pathology, which appears to be the ultimate cause of death in infected animals, is also consistently more severe in T1L than in T3D infected mice. Reovirus T1L and serotype 3 clone 9 (T3C9) can both reach the brain following peroral inoculation in CB17 SCID mice (George et al. 1990), although brain titers are low ($< 10^4$PFU/gm), and pathological lesions are apparently absent. It is important to recognize that infection in SCID mice differs in both the temporal profile and organ-specific patterns of injury from that seen following inoculation of neonatal mice from immunocompetent mouse strains.

In addition to showing longer survival than their T1L inoculated counterparts, SCID mice inoculated intraperitoneally with T3D generate organ-specific variants of T3D (Haller et al. 1995b). These organ specific variants include a brain-specific variant (T3DvBr) characterized by its capacity to grow faster in adult SCID mouse brain than wild-type T3D and to kill adult SCID mice more rapidly than wild-type T3D.

7 Genetics of CNS Tropism and Neurovirulence

7.1 Role of the S1 dsRNA Segment

The genetic basis for differences between the patterns of infection produced by the prototypic strains T1L and T3D have been analyzed using reassortant viruses (Weiner et al. 1977). Weiner et al. used eight T1L × T3D reassortants and found that the pattern of infection is determined by the S1 gene. Mice injected intracerebrally with reassortant viruses containing a T1L S1 gene survived and failed to develop neurological disease or developed hydrocephalus with minimal or absent neuronal injury. Mice injected with reassortants containing a T3D S1 gene developed lethal encephalitis without evidence of ependymitis or hydrocephalus (Weiner et al. 1977). These studies were subsequently extended using monoreassortants containing the T3D S1 gene on an otherwise completely T1L genetic background (1HA3) and the T1L S1 gene on an otherwise completely T3D genetic background (3HA1) (Weiner et al. 1980; Onodera et al. 1981).

Similar studies using T1L × T3D reassortants have expanded these original findings by indicating that the S1 dsRNA segment also determines the pattern of viral tropism and cell injury within the pituitary gland (Onodera et al. 1981) and retina (Tyler et al. 1985). The S1 dsRNA segment, in combination with the L1 and L2 dsRNA segments also plays a role in determining viral titer in the brains of adult SCID mice (Haller et al. 1995a) (see below).

7.1.1 T3D σ1 Monoclonal Antibody Resistant Variants

Attenuated reovirus T3D strains can be generated using neutralizing σ1 specific monoclonal antibodies (Spriggs and Fields 1982; Bassel-Duby et al. 1986). These variants show single nucleotide changes in the S1 dsRNA segment resulting in predicted amino acid substitutions at either position 340 or 419 within the globular head of the σ1 protein (Bassel-Duby et al. 1986). The growth pattern of these monoclonal antibody resistant (mar) T3D σ1 variants in L cells are generally identical to or only modestly lower than (e.g., peak titer reduced by tenfold for variant A) wild-type T3D (Spriggs and Fields 1982). The kinetics of replication of the T3D σ1 mar variants in L cells also does not appear to differ significantly from that of wild-type T3D (Spriggs and Fields 1982).

Following intracerebral inoculation into newborn mice T3D σ1 variants show strikingly attenuated growth and markedly reduced virulence compared to wild-type T3D (SPRIGGS and FIELDS 1982). Peak viral titers in brain generally range between 10^5 and 10^6PFU, nearly 10,000-fold less than that seen with wild-type virus (see above). The LD_{50} for these variants following intracerebral inoculation in newborn mice generally ranges between 10^5 and more than 10^7PFU, i.e., 10^3–10^6 times less virulent than wild-type T3D (SPRIGGS and FIELDS 1982; BASSEL-DUBY et al. 1986). In addition to altered growth and virulence these variants also show an altered pattern of CNS tissue injury (Fig. 10) (SPRIGGS et al. 1983; BASSEL-DUBY et al. 1986). Following intracerebral inoculation these variants produce striking injury to the hippocampus and mammillary bodies resulting in cavitary lesions, when brains are examined 15–21 days postinfection (SPRIGGS et al. 1983). Marked destructive lesions are also seen in septum and in parts of the hypothalamus and anterior colliculus (SPRIGGS et al. 1983).

Studies with a reassortant (T1L × T3D variant K) virus (1HAK) containing the S1 dsRNA segment derived from T3D variant K and all other dsRNA segments derived from T1L indicate that the variant K S1 dsRNA segment is responsible for the restricted pattern of CNS tissue injury, attenuated neurovirulence, and impaired CNS growth of variant K (KAYE et al. 1986).

7.2 Role of the M2 dsRNA Segment

Studies using reassortant viruses have also been used to determine the genetic basis for differences between more virulent (e.g., T3D, IC LD_{50} 10PFU/mouse) and less virulent (e.g., T3 H/Ta (now designated T3 clone 8), IC LD_{50} 10^4PFU/mouse) T3 strains (HRDY et al. 1982). The viral M2 gene, which encodes the major outer capsid protein μ1c, is the determinant of differences in lethality in T3 strains following intracerebral inoculation. The relative avirulence of the T3C8 strain is correlated with its diminished capacity to grow in brain, with peak brain titers of T3C8 being 100-fold lower than those seen with the virulent T3D strain (HRDY et al. 1982). Comparison of the nucleotide sequence of the T3D and T3C8 M2 genes and analysis of the predicted amino acid sequence of the encoded μ1 proteins indicate that there are 16 amino acid differences between the two proteins (JAYASURIYA 1991). These differences are spread throughout the protein, although several clustered near its hydrophilic externally facing carboxyterminal region (JAYASURIYA 1991). Interestingly, although the differences are not large, both the nucleotide sequence of the T3C8 M2 gene and the amino acid sequence of the encoded μ1 protein appears to be more closely related to those of T1L than to those of T3D (JAYASURIYA 1991).

Four serial passages of the relatively avirulent T3C8 through suckling mouse brain (T3C8B5) results in restoration of virulence approaching that of T3D (JAYASURIYA 1991; HRDY 1984). Reassortant studies suggest that the restored virulence of the brain-passaged isolate is determined by either the M1 or the M2 gene (HRDY 1984); however, sequence analysis of the M2 genes of T3C8 and T3C8B5 failed to reveal differences (JAYASURIYA 1991). This led to the suggestion that the M1 gene functions as an extragenic suppressor of the attenuated M2 gene in 8B5, allowing a restoration of virulence (JAYASURIYA 1991).

Additional suggestive evidence of a potential role for the M2 gene in neurovirulence comes from the study of ethanol-resistant mutants of T3D (WESSNER and FIELDS 1993). These mutants were derived from viruses surviving the treatment of a stock of T3D with 33% ethanol at 37°C for 20min. Genetic mapping studies indicated that mutations in the M2 gene are responsible for determining the ethanol resistant phenotype. Sequence analysis of seven clones revealed that six had mutations that would result in single amino acid substitutions at positions 319, 425, 440, and 454 of the μ1 protein, and one additional clone had two predicted amino acid substitutions at positions 233 and 442 (WESSNER and FIELDS 1993). These viruses are attenuated in neurovirulence and show lower growth in the CNS than do T3D (WESSNER 1991).

Perhaps surprisingly, the M2 gene segment is not implicated as a significant determinant of viral growth in the brain in adult SCID mice (HALLER et al. 1995a) (see below).

7.3 Genetics of Reovirus Growth in Brain of Adult SCID Mice

As noted above, T1L and T3D differ in the amount of virus present in the brains of SCID mice following intraperitoneal inoculation (HALLER et al. 1995a). Studies using T1L × T3D reassortants indicate that brain titers are determined by a combination of the L2 ($p = 0.005$), S1 ($p = 0.001$), and L1 ($p = 0.02$) genes (HALLER et al. 1995a). These genes encode the outer capsid protein σ1 and a small nonstructural protein σ1s (S1 gene), the core spike protein λ2 (L2 gene), and a minor core protein associated with RNA polymerase activity, λ3 (L1 gene). Linear regression analysis indicates that each of these genes contributes independently to determining viral titer in brain, and that taken together they account for 98% of the genetically determined variability in viral titer in brain. The proteins encoded by these genes are associated together as components of the viral vertex (HALLER et al. 1995a; see chapter by DERMODY, this volume). Oligomers of σ1 sit in a channel formed by pentamers of the core spike protein λ2. The core spike extends from the virion surface to the core. The base of the λ2 penton is in proximity to the minor core protein λ3. The exact mechanism by which the S1, L1, and L2 genes and their encoded proteins act to determine viral growth in brain remains unknown.

8 In Vitro Models of CNS Infection

8.1 Ependymal Cell Culture

Several cell culture model systems have been used to investigate reovirus infection of cells found within the CNS. Reovirus T1L, which infects ependymal cells in vivo, binds to the surface of both isolated human and murine ciliated ependymal cells (NEPOM et al. 1982; TARDIEU and WEINER 1982). Reovirus T3D, which does not produce significant ependymal cell infection in vivo, does not bind to ependymal cells in vitro (TARDIEU and WEINER 1982). The percentage of ependymal cells showing detectable T1L binding increases progressively from about 15% at a multiplicity of infection (MOI) of 20PFU (2000 virus particles) per cell to nearly 100% at an MOI of 10,000PFU (10^6 particles) per cell. Studies using the reovirus T1L × T3D monoreassortants 1HA3 and 3HA1 (see above) indicate that the T1L S1 gene determines the capacity of these viruses to bind to ependymal cells (TARDIEU and WEINER 1982), analogous to results found in vivo (WEINER et al. 1980). Unfortunately, studies of reovirus infection of ependymal cells in culture have been limited to binding studies and have not analyzed either viral replication or its capacity to produce cytopathic effects.

8.2 Neuronal Cell Culture

Reoviruses can also infect primary neuronal cell cultures derived from rat and mouse fetal cerebral cortex (NEPOM et al. 1982; DICHTER and WEINER 1984) and hippocampus (TYLER, unpublished). For studies of reovirus infection of cortical neurons (DICHTER and WEINER 1984) rat cortical neuronal cultures were established from embryonic day 15 (E15) rat cortices and plated onto collagen- and polylysine-coated dishes or coverslips. Early cultures (2- to 3-day-old) consist of neurons and neuronal precursor cells. Later cultures (3 weeks old) consist of a underlayer of glial and other nonneuronal cells and a more superficial layer of mature appearing neurons. These neurons stain with tetanus toxin and anti-neurofilament antibody and have the electrophysiological properties of neurons (DICHTER and WEINER 1984).

Both reovirus T1L and T3D replicate in both young and mature cultures, which, as noted, contain varying percentages of neurons, neuronal precursor cells, and nonneuronal cells. Peak titers achieved by T3D are approximately tenfold higher than those seen with T1L (DICHTER and WEINER 1984). Cultures show cytopathic effects (CPE) as early as 24h postinfection, which are typically maximal by 5 days postinfection (DICHTER and WEINER 1984). Immunofluorescent localization of T1L infection indicates that this virus infects primarily the nonneuronal cells, including astrocytes, rather than the neuronal cells present in these cultures. By contrast, T3D infects the neuronal cells and produces a variable degree of infection in the nonneuronal cell population. Viral antigen can be seen in both the cell body and processes of infected neurons. Not all neurons within these cultures are infected even at high MOI, suggesting that there may be some intrinsic variability to neuronal susceptibility to reovirus infection. Binding studies using T1L, T3D, and the T1L × T3D monoreassortants 1HA3 and 3HA1, indicate that the T3D S1 gene determines the capacity of these viruses to bind to neuronal cells in these cultures (DICHTER and WEINER 1984). Neither T1L nor 3HA1 shows significant binding to neuronal cells.

Reovirus infection of neuronal cultures has also been studied using a syngeneic anti-idiotypic antibody (87.92.6) generated against the reovirus T3D-σ1 specific monoclonal antibody 9BG5. Sequence analysis of the complementarity-determining region of the variable portion of the light chain of this antibody show that there is a 17 amino acid long region of sequence similarity (internal image) between this antibody and a portion (amino acids 317–332) of the globular head of the T3D σ1 protein. Studies using this antibody as a surrogate marker for σ1 binding in neuronal cultures suggest that the majority (> 90%) of neuronal cells and neuronal precursors in both early and late cultures derived from rat fetal cortices have receptors for reovirus T3D (DICHTER et al. 1986). Pretreatment of neuronal cultures with 87.92.6 reduces the number of T3D infected neurons by 75% but does not significantly affect T3D infection of nonneuronal cells present in the cultures (DICHTER et al. 1986).

8.3 Pituitary Cell Culture

Differential affects of reovirus infection can also be seen in cultured pituitary cell lines (MARATOS-FLIER et al. 1983, 1985). Scatchard analysis indicates that the rat pituitary adenoma cell line GH_4C_1 contains 4200 viral binding sites per cell, with an apparent binding affinity for reovirus of 1.2×10^{-11} M (MARATOS-FLIER et al. 1983). Both reovirus T1L and T3D grow to equivalent titers in the GH_4C_1 rat pituitary adenoma cell line, achieving near maximal titers 24–48h postinfection. Despite the similar growth rates for T1L and T3D in GH_4C_1 cells, there are striking differences in CPE between these two viral strains. At an MOI of 4, by 72h postinfection 60% of cells in T3D-infected cultures are dead, compared to only 20% of T1L-infected cells. The degree of CPE is dependent on the size of the viral inoculum. For example, at an MOI of 50 nearly 75% of T3D infected cells are dead within 48h of infection. However, regardless of the MOI employed T3D is significantly more cytopathic than T1L. Studies of macromolecular synthesis in these cells suggest that differences in the capacity of these viruses to alter macromolecular synthesis in these cells do not explain differences in their CPE. However, T3D does appear to result in more inhibition of DNA synthesis 0–48h postinfection than T1L. T3D-infected cells also show slightly lower levels of protein synthesis than their T1L-infected counterparts. These differences are relatively modest and in some cases are not statistically significant (MARATOS-FLIER et al. 1985).

8.4 Oligodendrocyte Culture

8.4.1 Effects of Reovirus Infection on Cellular Differentiation Programs

Reovirus infection of glial progenitor (O-2A) cells derived from neonatal rat optic nerve alters the development program of these cells (COHEN et al. 1990, 1991). O-2A glial progenitor cells differentiate into oligodendrocytes and type 2 (fibrous) astrocytes. This differentiation program can be followed with the use of monoclonal antibodies to cell-type specific markers. For example, the presence of galactocerebroside and myelin basic protein serve as markers for mature oligodendrocytes. Reovirus receptor does not appear to be present on immature O-2A progenitor cells, but is detectable on the vast majority of cells that are either galactocerebroside or myelin basic protein positive. Treatment of O-2A cultures with the 87.92.6 anti-idiotypic antibody, which (see above) has a region of sequence similarity to the T3D σ1 cell attachment protein, or with synthetic peptides corresponding to the region in question, induces premature appearance of galactocerebroside and myelin basic protein in O-2A cultures (COHEN et al. 1990, 1991). As a result of these studies it was proposed that cell surface receptors used as attachment sites by T3D also play a role in oligodendrocyte differentiation (COHEN et al. 1990, 1991).

8.4.2 Effects of Reovirus Infection on Oligodendrocytes and Myelin In Vivo

The 87.92.6 antibody may also alter oligodendrocyte function in vivo (COHEN et al. 1992). Microinjection of purified 87.92.6 antibody into guinea pig optic nerve in vivo produces alterations in oligodendrocyte and myelin morphology. Following injection myelinated optic nerve fibers show separation of myelin lamellae, myelin vesiculation, widening of the Schmidt-Lanterman clefts, and regions of demyelination (COHEN et al. 1992). Prominent myelin changes are visible within 4h of antibody microinjection, appear in up to 25% of myelinated axons in the area of injection, and appear distinct from changes induced by microinjection of control solutions. Up to 10% of axons within the center of the microinjected region show complete demyelination, with preservation of axonal morphology. These effects do not appear to be the result of either local interstitial edema or an inflammatory cell response and can be abrogated by blocking the antigen-binding domain of 87.92.6 prior to injection. Similar effects to those described with 87.92.6 were also seen with monovalent and divalent synthetic peptides containing the 17 amino acid region of 87.92.6 with sequence similarity to amino acids 317–332 of the T3D σ1 protein (COHEN et al. 1992). Although these results indicate that under certain circumstances perturbation of the T3D receptor in vivo may have direct myelinotoxic effects, it is important to emphasize that demyelination has never been described as a feature of either natural or experimental reovirus infection of the CNS.

References

Amerongen HM, Weltzin RA, Farnet CM, Michetti P, Haseltine WA, Neutra MR (1991) Transepithelial transport of HIV by intestinal M cells: a mechanism for transmission of AIDS. J Acquir Immune Defic Syndr Hum Retrovirol 4:760–765

Amerongen HM, Wilson GAR, Fields BN, Neutra MR (1994) Proteolytic processing of reovirus is required for adherence to intestinal M cells. J Virol 68:8428–8432

Armstrong GD, Paul RW, Lee PW (1984) Studies on reovirus receptors of L cells: virus binding characteristics and comparison with reovirus receptors of erythrocytes. Virology 138:37–48
Averback P (1982) Reovirus and pathogenesis of some forms of chronic mental illness. Med Hypotheses 8:383–391
Babiss LE, Luftig RB, Weatherbee JA, Weihing RR, Ray UR, Fields BN (1979) Reovirus serotypes 1 and 3 differ in their in vitro association with microtubules. J Virol 30:863–874
Barthold SW, Smith AL, Bhatt PN (1993) Infectivity, disease patterns, and serologic profiles of reovirus serotypes 1, 2, and 3 in infant and weanling mice. Lab Anim Sci 43:425–430
Bass DM, Trier JS, Dambrauskas R, Wolf JL (1988) Reovirus type 1 infection of small intestinal epithelium in suckling mice and its effect on M cells. Lab Invest 58:226–235
Bass DM, Bodkin D, Dambrauskas R, Trier JS, Fields BN, Wolf JL (1990) Intraluminal proteolytic activation plays an important role in replication of type 1 reovirus in the intestines of neonatal mice. J Virol 64:1830–1833
Bássel-Duby R, Spriggs DR, Tyler KL, Fields BN (1986) Identification of attenuating mutations on the reovirus type 3 S1 double-stranded RNA segment with a rapid sequencing technique. J Virol 60:64–67
Chappell JD, Gunn VL, Wetzel JD, Baer GS, Dermody TS (1997) Mutations in type 3 reovirus that determine binding to sialic acid are contained in the fibrous tail domain of viral attachment protein sigma 1. J Virol 71:1834–1841
Choi AH, Lee PWK (1988) Does the beta-adrenergic receptor function as a reovirus receptor? Virology 163:191–197
Choi AH, Paul RW, Lee PW (1990) Reovirus binds to multiple plasma membrane proteins of mouse L fibroblasts. Virology 178:316–320
Co MS, Gaulton GN, Fields BN, Greene MI (1985a) Isolation and biochemical characterization of the mammalian reovirus type 3 cell-surface receptor. Proc Natl Acad Sci USA 82:1494–1498
Co MS, Gaulton GN, Tominaga A, Homcy CJ, Fields BN, Greene MI (1985b) Structural similarities between the mammalian beta-adrenergic and reovirus type 3 receptors. Proc Natl Acad Sci USA 82:5315–5318
Cohen JA, Williams WV, Weiner DB, Geller HM, Greene MI (1990) Ligand binding to the cell surface receptor for reovirus type 3 stimulates galactocerebroside expression by developing oligodendrocytes. Proc Natl Acad Sci USA 87:4922–4926
Cohen JA, Williams WV, Geller HM, Greene MI (1991) Anti-reovirus receptor antibody accelerates expression of the optic nerve oligodendrocyte developmental program. Proc Natl Acad Sci USA 88:1266–1270
Cohen JA, Sergott RC, Williams WV, Hill SJ, Brown MJ, Greene MI (1992) In vivo modulation of oligodendrocyte function by an anti-receptor antibody. Pathobiology 60:151–156
Csiza CK (1974) Characterization and serotyping of three feline reovirus isolates. Infect Immun 9: 159–166
Davis LE (1982) Experimental viral infection of the inner ear. III. Viremic spread of reovirus to hamster eighth nerve ganglion cells. Ann Otol Rhinol Laryngol 91:90–93
Dermody TS, Nibert ML, Bassel-Duby R, Fields BN (1990) A sigma 1 region important for hemagglutination by serotype 3 reovirus strains. J Virol 64:5173–5176
Dichter MA, Weiner HL (1984) Infection of neuronal cell cultures with reovirus mimics in vitro patterns of neurotropism. Ann Neurol 16:603–610
Dichter MA, Weiner HL, Fields BN, Mitchell G, Noseworthy J, Gaulton GN et al (1986) Antiidiotypic antibody to reovirus binds to neurons and protects from viral infection. Ann Neurol 19:555–558
Donta ST, Shanley JD (1990) Reovirus type 3 binds to antagonist domains of the beta-adrenergic receptor. J Virol 64:639–641
Flamand A, Gagner JP, Morrison LA, Fields BN (1991) Penetration of the nervous systems of suckling mice by mammalian reoviruses. J Virol 65:123–131
Gentsch JR, Hatfield JW (1984) Saturable attachment sites for type 3 mammalian reoviruses on murine L cells and human HeLa cells. Virus Res 1:401–424
Gentsch JR, Pacitti AF (1985) Effect of neuraminidase treatment of cells and effects of soluble glycoproteins on type 3 reovirus attachment to murine L cells. J Virol 54:307–315
Gentsch JR, Pacitti AF (1987) Differential interaction of reovirus type 3 with sialylated receptor component on animal cells. Virology 161:245–248
George A, Kost SI, Witzleben CL, Cebra JJ, Rubin DH (1990) Reovirus-induced liver disease in severe combined immunodeficient (SCID) mice. J Exp Med 171:929–934
Gonatas NK, Margolis G, Kilham L (1971) Reovirus type III encephalitis: observations of virus-cell interactions in neural tissues. II. Electron microscopic studies. Lab Invest 24:101–109

Haller BL, Barkon ML, Vogler GP, Virgin HW (1995a) Genetic mapping of reovirus virulence and organ tropism in severe combined immunodeficient mice: organ-specific virulence genes. J Virol 69:357–364
Haller BL, Barkon ML, Li X-Y, Hu WM, Wetzel JD, Dermody TS et al (1995b) Brain- and intestine-specific variants of reovirus serotype 3 strain Dearing are selected during chronic infection of severe combined immunodeficient mice. J Virol 69:3933–3937
Hrdy DB (1984) Virus infection and primate evolution with special reference to the reoviridae. Thesis, Harvard University
Hrdy DB, Rubin DN, Fields BN (1982) Molecular basis of reovirus neurovirulence: role of the M2 gene in avirulence. Proc Natl Acad Sci USA 79:1298–1302
Jayasuriya AK (1991) Molecular characterization of the reovirus M2 gene. Thesis, Harvard University
Jenson AB, Rabin ER, Phillips CA, Melnick JL (1965) Reovirus encephalitis in newborn mice: an electron microscopic and virus assay study. Am J Pathol 47:223–239
Johansson PJH, Sveger T, Ahlfors K, Ekstrand J, Svensson L (1996) Reovirus type 1 associated with meningitis. Scand J Infect Dis 28:117–120
Joske RA, Keall DD, Leak PJ, Stanley NF, Walters MN (1964) Hepatitis-encephalitis in humans with reovirus infections. Arch Intern Med 113:811–816
Kauffman RS, Wolf JL, Finberg R, Trier JS, Fields BN (1983) The sigma 1 protein determines the extent of spread of reovirus from the gastrointestinal tract of mice. Virology 124:403–410
Kaye KM, Spriggs DR, Bassel-Duby R, Fields BN, Tyler KL (1986) Genetic basis for altered pathogenesis of an immune-selected antigenic variant of reovirus type 3 Dearing. J Virol 59:90–97
Kilham L, Margolis G (1969) Hydrocephalus in hamsters, ferrets, rats and mice following inoculations with reovirus type 1. I. Virologic studies. Lab Invest 21:183–188
Krainer L, Aronson BE (1959) Disseminated encephalomyelitis in humans with recovery of hepatoencephalitis virus. J Neuropathol Exp Neurol 18:339–342
Kundin WD, Liu C, Gigstad J (1966) Reovirus infection in suckling mice: immunofluorescent and infectivity studies. J Immunol 97:393–401
Liu J, Co MS, Greene MI (1988) Reovirus type 3 and [125I]-iodocyanopindolol bind to distinct domains on the beta-adrenergic like receptor. Immunol Res 7:232–238
Maratos-Flier E, Kahn CR, Spriggs DR, Fields BN (1983) Specific plasma membrane receptors for reovirus on rat pituitary cells in culture. J Clin Invest 72:617–621
Maratos-Flier E, Goodman MJ, Fields BN, Kahn CR (1985) Differential effects of viral infection on islet and pituitary cell lines. Endocrinology 116:2430–2437
Margolis G, Kilham L (1969a) Hydrocephalus in hamsters, ferrets, rats, and mice following inoculations with reovirus type 1. II. Pathologic studies. Lab Invest 21:189–198
Margolis G, Kilham L (1969b) Experimental virus-induced hydrocephalus. Relation to pathogenesis of the Arnold-Chiari malformation. J Neurosurgery 31:1–9
Margolis G, Kilham L, Gonatos N (1971) Reovirus type III encephalitis: observations of virus-cell interactions in neural tissues. I. Light microscopy studies. Lab Invest 24:91–109
Massie EL, Shaw ED (1966) Reovirus type 1 in laboratory dogs. Am J Vet Res 27:783–878
Masters C, Alpers M, Kakulas B (1977) Pathogenesis of reovirus type 1 hydrocephalus in mice: significance of aqueductal changes. Arch Neurol 34:18–28
Milhorat TH, Kotzen RM (1994) Stenosis of the central canal of the spinal cord following inoculation of suckling hamsters with reovirus type 1. J Neurosurgery 81:103–106
Morin MJ, Warner A, Fields BN (1994) A pathway of entry of reoviruses into the host through M cells of the respiratory tract. J Exp Med 180:1523–1527
Morin MJ, Warner A, Fields BN (1996) Reovirus infection in rat lungs as a model to study the pathogenesis of viral pneumonia. J Virol 70:541–548
Morrison LA, Fields BN (1991) Parallel mechanisms in neuropathogenesis of enteric virus infections. J Virol 65:2767–2772
Morrison LA, Sidman RL, Fields BN (1991) Direct spread of reovirus from the intestinal lumen to the central nervous system through vagal autonomic nerve fibers. Proc Natl Acad Sci USA 88:3852–3856
Morrison LA, Fields BN, Dermody TS (1993) Prolonged replication in the mouse central nervous system of reoviruses isolated from persistently infected cell cultures. J Virol 67:3019–3026
Nepom J, Weiner HL, Dichter MA, Tardieu M, Spriggs DR, Gram M, Powers ML, Fields BN, Greene MI (1982) Identification of a hemagglutinin-specific idiotype associated with reovirus recognition shared by lymphoid and neural cells. J Exp Med 155:155–167
Neutra MR, Kraehenbuhl JP (1992) Transepithelial transport and mucosal defense. I. The role of M cells. Trends Cell Biol 2:134–138

Nielsen SL, Baringer JR (1972) Reovirus-induced aqueductal stenosis in hamsters: phase contrast and electron microscopic studies. Lab Invest 27:531–537

Oberhaus SM, Smith RL, Clayton GH, Dermody TS, Tyler KL (1997) Reovirus infection and tissue injury in the mouse central nervous system are associated with apoptosis. J Virol 71:2100–2106

Onodera T, Toniolo A, Ray UR, Jenson AB, Knazek RA, Notkins AL (1981) Virus-induced diabetes mellitus. XX. Polyendocrinopathy and autoimmunity. J Exp Med 153:1457–1473

Papadimitriou JM (1967) An electron microscopic study of murine reovirus 3 encephalitis. Am J Pathol 50:59–75

Paul RW, Lee PWK (1987) Glycophorin is the reovirus receptor on human erythrocytes. Virology 159:94–101

Paul RW, Choi AH, Lee PWK (1989) The alpha-anomeric form of sialic acid is the minimal receptor determinant recognized by reovirus. Virology 172:382–385

Phillips PA, Alpers MP, Stanley NF (1970) Hydrocephalus in mice inoculated neonatally by the oronasal route with reovirus type 1. Science 168:858–859

Raine CS, Fields BN (1973) Ultrastructural features of reovirus type 3 encephalitis. J Neuropathol Exp Neurol 32:19–33

Rubin DH, Wetzel JD, Williams WV, Cohen JA, Dworkin C, Dermody TS (1992) Binding of type 3 reovirus by a domain of the sigma 1 protein important for hemagglutination leads to infection of murine erythroleukemia cells. J Clin Invest 90:2536–2542

Sabin AB (1959) Reoviruses. Science 130:1387–1389

Sawutz DG, Bassel-Duby R, Homcy CJ (1987) High affinity binding of reovirus type 3 to cells that lack beta adrenergic receptor activity. Life Sci 40:399–406

Sicinski P, Rowinski J, Warchol JB, Jarzcabek Z, Gut W, Szczygiel B et al (1990) Poliovirus type 1 enters the human host through intestinal M cells. Gastroenterology 98:56–58

Spriggs DR, Fields BN (1982) Attenuated type 3 strains generated by selection of hemagglutinin antigenic variants. Nature 297:68–70

Spriggs DR, Bronson RT, Fields BN (1983) Hemagglutinin variants of reovirus type 3 have altered central nervous system tropism. Science 220:505–507

Stanley NF, Dorman DC, Ponsford J (1953) Studies on the pathogenesis of a hitherto undescribed virus (hepato-encephalomyelitis) producing unusual symptoms in suckling mice. Aust J Exp Biol 31: 147–160

Stanley NF, Dorman DC, Ponsford J (1954) Studies on the hepato-encephalomyelitis virus (HEV). Aust J Exp Biol 32:543–562

Stanley NF, Leak PJ, Walters MN, Joske RA (1964) Murine infection with reovirus. II. The chronic disease following reovirus type 3 infection. Br J Exp Pathol 45:142–149

Strong JE, Tang D, Lee PWK (1993) Evidence that the epidermal growth factor receptor on host cells confers reovirus infection efficiency. Virology 197:405–411

Szirmai I, Antalicz M, Trombitas K, Kuntar L, Gati I (1983) Adult onset rapidly progressive spinal muscle atrophy of shoulder girdle with gammopathy. Clinical Neuropathol 2:128–133

Tang D, Strong JE, Lee PWK (1993) Recognition of the epidermal growth factor receptor by reovirus. Virology 197:412–414

Tardieu M, Weiner HL (1982) Viral receptors on isolated murine and human ependymal cells. Science 215:419–421

Tardieu M, Powers ML, Weiner HL (1983) Age-dependent susceptibility to reovirus type 3 encephalitis: role of viral and host factors. Ann Neurol 13:602–607

Tyler KL (1991) Pathogenesis of reovirus infections of the central nervous system. Semin Neurosci 3: 117–124

Tyler KL, Fields BN (1988) Reovirus infections of the central nervous system. In: Johnson RT, Lyon G (eds) Virus infections of the developing nervous system. Kluwer, Dordrecht, pp 11–19

Tyler KL, Fields BN (1996) Reoviruses. In: Fields BN, Knipe DM (eds) Fields' virology. Raven, New York, pp 1597–1623

Tyler KL, Bronson RT, Byers KB, Fields BN (1985) Molecular basis of viral neurotropism: experimental reovirus infection. Neurology 35:88–92

Tyler KL, McPhee DA, Fields BN (1986) Distinct pathways of viral spread in the host determined by reovirus S1 gene segment. Science 233:770–774

Tyler KL, Virgin HW IV, Bassel-Duby R, Fields BN (1989) Antibody inhibits defined stages in the pathogenesis of reovirus serotype 3 infection of the central nervous system. J Exp Med 170:887–900

Tyler KL, Mann MA, Fields BN, Virgin HW IV (1993) Protective anti-reovirus monoclonal antibodies and their effects on viral pathogenesis. J Virol 67:3446–3453

Van Tongeren HAE (1957) A familial infection with hepatoencephalomyelitis virus in the Netherlands: study on some properties of the infective agent. Arch Gesamt Virusforsch 7:429–448

Virgin HW IV, Bassel-Duby R, Fields BN, Tyler KL (1988) Antibody protects against lethal infection with the neurally spreading Reovirus type 3 (Dearing). J Virol 62:4594–4604

Virgin HW IV, Tyler KL, Dermody TS (1997) Reovirus. In: Nathanson N (ed) Viral pathogenesis. Lippincott-Raven, New York, pp 669–699

Walters MN, Joske RA, Leak PJ, Stanley NF (1963) Murine infection with reovirus. I. Pathology of the acute phase. Br J Exp Pathol 44:427–436

Walters MN, Leak PJ, Joske RA, Stanley NF, Perret DH (1965) Murine infection with reovirus. III. Pathology of infection with types 1 and 2. Br J Exp Pathol 46:200–212

Weiner HL, Drayna D, Averill DRJR, Fields BN (1977) Molecular basis of reovirus virulence: role of the S1 gene. Proc Natl Acad Sci USA 74:5744–5748

Weiner HL, Powers ML, Fields BN (1980) Absolute linkage of virulence and central nervous system tropism of reoviruses to viral hemagglutinin. J Infect Dis 141:609–616

Wessner DR (1991) Isolation and characterization of ethanol resistant mutants of reovirus. Thesis, Harvard University

Wessner DR, Fields BN (1993) Isolation and genetic characterization of ethanol-resistant reovirus mutants. J Virol 67:2442–2447

Wolf JL, Rubin DH, Finberg R, Kauffman RS, Sharpe AH, Trier JS et al (1981) Intestinal M cells: a pathway of entry of reovirus into the host. Science 212:471–472

Wolf JL, Kauffman RS, Finberg R, Dambrauskas R, Fields BN (1983) Determinants of reovirus interaction with intestinal M cells and absorptive cells of murine intestine. Gastroenterology 85:291–300

Wolf JL, Dambrauskas R, Sharpe AH, Trier JS (1987) Adherence to and penetration of the intestinal epithelium by reovirus type 1 in neonatal mice. Gastroenterology 92:82–91

Reoviruses and the Interferon System

C.E. SAMUEL

1 Introduction

1.1 The Interferon System

What are interferons (IFNs)? How do IFNs function to reduce the efficiency of reovirus replication? Considerable progress has been made toward answering these and many other important questions concerning the IFN system. In the course of answering these questions fundamental new insights have been gained that advance our understanding of the processes used to control gene expression in animal cells.

Interferon was discovered as an antiviral agent during studies on virus interference. ISAACS and LINDENMANN (1957) demonstrated that influenza-infected chick cells produce a secreted factor which mediates the transfer of a virus-resistant state that is functional against both homologous and heterologous viruses. This fundamental observation, and similar findings by NAGONO and KOJIMA (1958), set the stage for subsequent studies by many laboratories around the world that led to the elucidation of the IFN system in exquisite detail. We now know that IFNs are members of the larger cytokine family of proteins, and that they possess a wide

Department of Molecular, Cellular and Developmental Biology, and Interdepartmental Graduate Program of Biochemistry and Molecular Biology, University of California, Santa Barbara, CA 93106-9610, USA
(E-mail: samuel@lifesci.lscf.ucsb.edu)

range of biological activities in addition to their characteristic antiviral activity by which they were discovered (VILCEK and SEN 1996). The IFN system includes those cells that synthesize IFN in response to an external stimulus such as viral infection, and those cells that respond to IFN by establishing an antiviral state (PESTKA et al. 1987; SAMUEL 1991). Many animal viruses, including reoviruses, are both inducers of IFNs and are sensitive to the antiviral actions of IFNs. Some of the key features of the IFN system are summarized in Fig. 1.

IFNs are a multigene family of cytokines (DIAZ et al. 1993; BLATT et al. 1996; YOUNG 1996) that are commonly grouped into two types, type I IFN and type II IFN (SAMUEL 1991). Type I IFNs, often referred to as viral IFNs, include four subtypes: IFN-α (αI, leukocyte), IFN-β (fibroblast), IFN-ω (αII), and IFN-τ (trophoblast). Type II IFN is frequently referred to as immune IFN or IFN-γ. The type I IFNs are induced by virus infection, whereas type II IFN is induced by

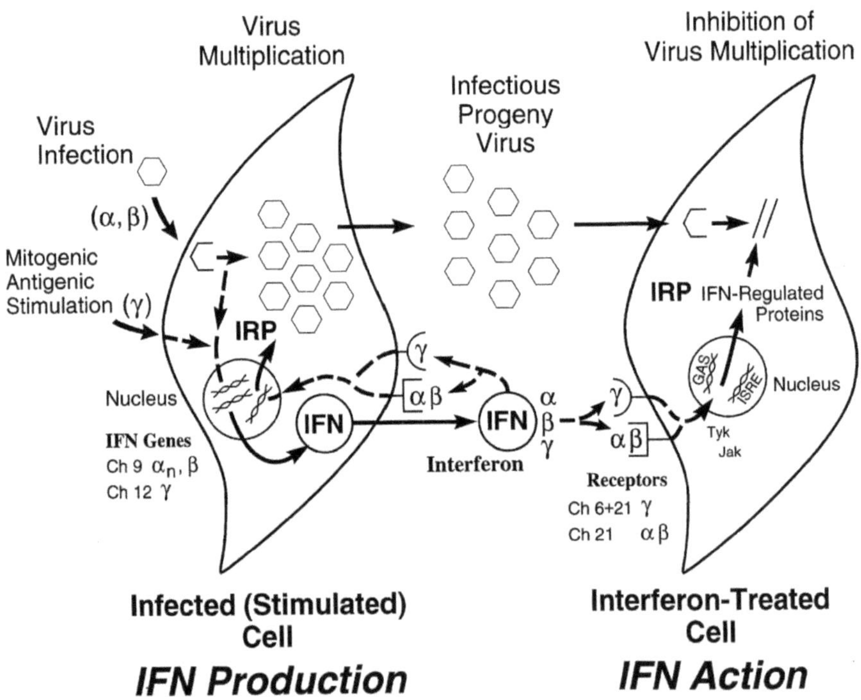

Fig. 1. A schematic summary diagram of the IFN system. *Open hexagons*, virion particles; *open circles*, IFN proteins. *Left*, an IFN-producing cell depicting a cell induced to synthesize IFN in response to either virus infection (IFN-α and IFN-β) or antigens or mitogen stimulation (IFN-γ). *Right*, an IFN-treated cell depicting a cell induced to synthesize IFN-regulated proteins that collectively constitute the antiviral response responsible for the inhibition of virus multiplication. Among the IFN-induced dsRNA-binding proteins that may contribute to the reduced multiplication of viruses, including reoviruses, within single cells are PKR that inhibits translation initiation, the 2′,5′-oligoadenylate synthetase-RNase that mediates RNA degradation, and the ADAR adenosine deaminase that edits RNA. IFN-induced expression of MHC class I and II antigens may contribute to the antiviral responses observed within whole animals. *Numerals*, the chromosome assignment of the human genes encoding the IFNs and their receptors

mitogenic or antigenic stimuli. Most types of cells are capable of production of the type I α/β IFNs; type II IFN-γ is synthesized by certain cells of the immune system including CD4 TH1 cells, CD8 cytotoxic suppressor cells, and natural killer cells. The large number of IFN-α genes, and the single IFN-β gene, all lack introns and are clustered on the short arm of chromosome 9 in the human, and chromosome 4 in the mouse. The single IFN-γ gene possesses three introns and maps to the long arm of chromosome 12 in the human, and chromosome 10 in the mouse. Although some IFNs are modified posttranslationally by N- and O-glycosylation, the major human type I IFN-α subspecies are not glycosylated. The IFN-α's appear to function as monomers, whereas IFN-β and IFN-γ appear to function as homodimers.

Both type I and type II IFNs possess antiviral and immunoregulatory activities (PESTKA et al. 1987). IFNs exert their actions through species-specific cell surface receptors. Type I IFNs, the α's and β, appear to share a common receptor consisting of at least two subunits, with both components mapping to chromosome 21 in the human, and chromosome 16 in the mouse. Type II IFN-γ binds to a receptor that is distinct from that used by IFN-α and β. The two components of the IFN-γ receptor map to chromosomes 6 and 21 in the human, and chromosomes 10 and 16 in the mouse. Characterization of the ability of IFNs to induce an antiviral state led not only to the identification of several IFN-induced proteins responsible for their actions (SAMUEL 1991; SEN and RANSOHOFF 1992) but also to the elucidation of the mechanisms of signal transduction and transcriptional activation of the cellular genes encoding the IFN-induced proteins (DARNELL et al. 1994; SCHINDLER and DARNELL 1995).

IFN-mediated signaling and transcriptional activation occurs via a pathway commonly known as the Jak-STAT pathway. Receptor-associated Janus tyrosine kinases are activated following binding of IFN to the cognate R1/R2 multicomponent transmembrane receptor. Signaling by the two types of IFNs involves overlapping subsets of kinases; Jak1 and Tyk2 kinases for the type I IFNs, and the Jak1 and Jak2 kinases for type II IFN. Activation of the receptor-associated kinases leads to the subsequent phosphorylation of latent cytoplasmic STAT transcription factors. For IFN-α and -β IFNs, the phosphorylated Stat1α/β and Stat2 factors along with an additional non-STAT protein, p48, translocate to the nucleus and form a complex known as ISGF3α. ISGF3α binds to a *cis*-acting element, designated ISRE, commonly found in type I IFN inducible genes. For IFN-γ, the phosphorylated Stat1α factor homodimerizes, translocates to the nucleus and binds to a different *cis*-acting element, designated GAS, that is commonly found in type II IFN inducible genes. Several cellular proteins, whose syntheses are induced by IFN, act individually or synergistically to establish the IFN-induced antiviral state. Some of these proteins, for example, the RNA-dependent protein kinase (PKR) (SAMUEL 1993), appear to play a central role in the inhibition of reovirus multiplication in IFN-treated cells. Other IFN-induced proteins, such as the 2′-5′ oligoadenylate synthetase-nuclease (WILLIAMS and SILVERMAN 1985) and protein Mx (STAEHELI 1990), are implicated in the inhibition of picornaviruses and negative-stranded viruses, respectively, in IFN-treated cells (SAMUEL 1991).

1.2 The Molecular Biology of Reovirus

The genetics and biochemistry of the human reoviruses are reviewed in depth in accompanying chapters of this volume (see, e.g., the chapter by K.L. TYLER, this volume). One of the principle features of the human reoviruses, members of the family of reoviridae (JOKLIK 1983), is that they possess a segmented, double-stranded RNA (dsRNA) genome. Indeed, it is this property of a segmented genome that Fields and his students and fellows so eloquently exploited, generating genetic reassortant hybrid reoviruses. These reassortant viruses were used in subsequent seminal studies to elucidate functions of individual reovirus genome products and to define aspects of the molecular basis of reovirus pathogenesis (FIELDS and GREENE 1982). The principle features of the reovirus multiplication cycle essential to our understanding of the use of the virus in studies of the IFN system are summarized in Fig. 2.

The genome of human reovirus consists of ten segments of dsRNA, each of which is transcribed into single-stranded RNA of the polarity of mRNA by a virion-associated dsRNA-dependent ssRNA polymerase. The reovirus mRNA transcripts, nine of which are monocistronic, encode twelve protein products. The s1 mRNA transcribed from the S1 dsRNA segment is bicistronic and encodes two proteins, the minor capsid attachment protein σ1 and the nonstructural protein σ1NS. The major outer protein μ1C is derived by posttranslational proteolytic cleavage from the μ1 product of the M2 dsRNA segment. Primary transcription of the segmented reovirus genome catalyzed by the virion-associated polymerase takes place in the cytoplasm of the infected host and occurs in the absence of protein synthesis. Primary gene expression from parental particles typically accounts for less than 5% of the total viral RNA and protein synthesized in an infected cell. Viral protein synthesis is required for synthesis of viral dsRNA, a process that occurs during particle morphogenesis. Secondary gene expression from progeny core-like particles accounts for the vast majority of the viral mRNA, and thus viral protein, synthesized in cells productively infected with reovirus.

2 Induction of Interferon by Reovirus

Interferon is an inducible cytokine, and viral infection is the most common means of induction (Fig. 1). Reovirus infection of animal cells can lead to the induction of IFN. The IFN inducing capacity of reovirus is dependent both upon the particular reovirus strain and the kind of cell or host animal that is infected. For example, in mouse fibroblasts in culture, wild-type serotype 3 reovirus is a better IFN inducer than serotype 1 reovirus (SHARPE and FIELDS 1983). However, all serotypes induce IFN. In primary human thyroid follicular cells, both serotypes 1 and 3 of reovirus induce type I IFN-α and -β production as measured by enhanced MHC class I antigen expression (ATTA et al. 1995). The enhanced MHC I expression induced by

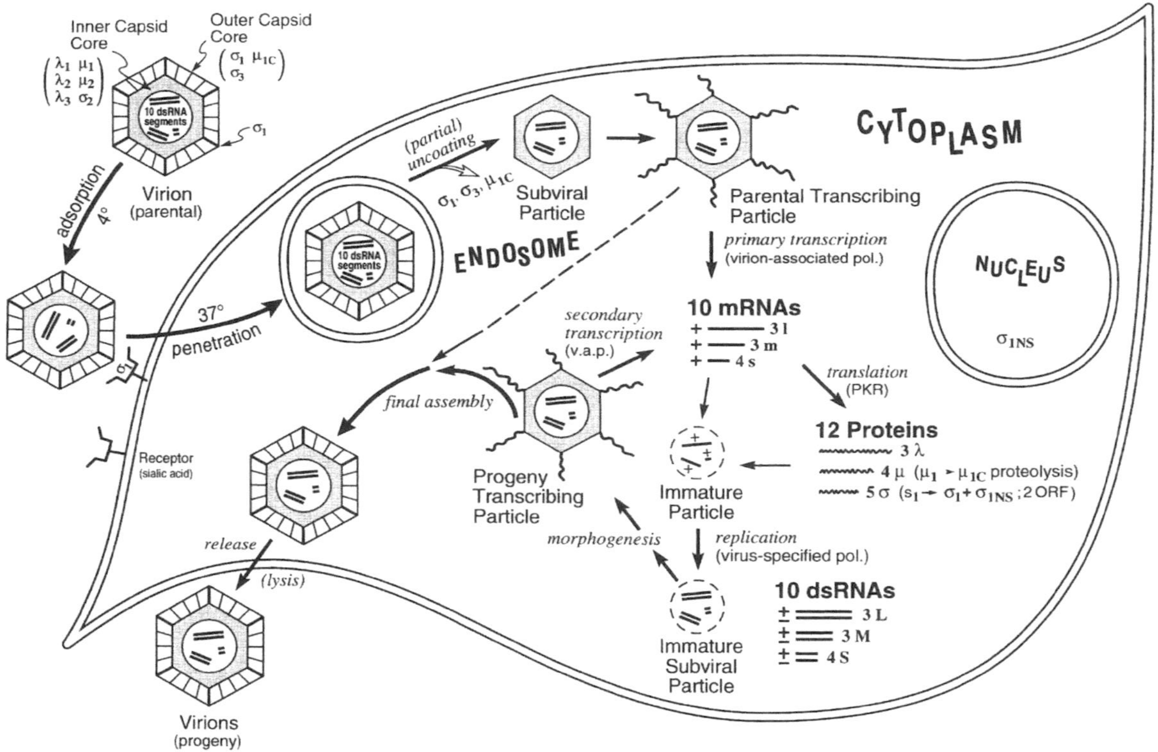

Fig. 2. A schematic diagram of the reovirus multiplication cycle. *Hexagons*, reovirion particles; *parallel lines*, the ten dsRNA genome segments that fall into three-sized classes (3L, 3M, 4S); *single lines*, the ten (3*l*, 3*m*, 4*s*) single-stranded plus-polarity mRNA transcripts produced by the virion-associated RNA polymerase. The 12 known virus-encoded proteins (3 λ, 4 μ, 5 σ) include the 9 structural proteins that constitute the virion inner ($\lambda 1$, $\lambda 2$, $\lambda 3$, $\mu 1$, $\mu 2$, $\sigma 2$) and outer ($\mu 1C$, $\sigma 1$, $\sigma 3$) capsid shells, and three nonstructural proteins (μNS, σNS, $\sigma 1NS$) found only within infected cells and not part of the virion particle. Multiplication occurs within the cytoplasm of the infected cell. *PKR*, the IFN-inducible RNA-dependent protein kinase, is important in the control of translation in virus-infected, IFN-treated cells

reovirus infection in thyroid donor cultures is neutralized by antibody against IFN-α and IFN-β.

Infection of rat insulinoma RIN-m5F cells with either serotype 1 or serotype 3 also induces MHC class I expression via induction of IFN (CAMPBELL and HARRISON 1989). Reovirus infection also induces MHC class II expression in Fisher rat 1B-6 thyroid cells; serotype 3 is a somewhat more efficient inducer than serotype 1 (NEUFELD et al. 1989). Type II IFN-γ production can also be induced by reovirus infection in whole animals. The capacity of serotype 1 Lang strain virus to induce IFN-γ in peripheral lymph node lymphocytes is dependent upon the mouse strain; lymphocytes from C3H mice produce significantly higher levels of type I IFN-γ than do those from BALB/c, C57BL/6, and B10.D2 mice (MAJOR and CUFF 1996).

So far it has not been possible to definitively identify which reovirus genes, or multiplication events, primarily determine the IFN-inducing capacity of the virus. The IFN-inducing capacity of prototype serotype 3 Dearing strain temperature-sensitive mutants from six complementation groups (tsA, tsB, tsC, tsD, tsE, tsG), when examined using mouse L929 fibroblasts in culture, is comparable to that of wild-type parental virus at 31°C, a permissive temperature for replication. By contrast, at 38.5°C, a nonpermissive temperature for virus replication, all of the ts mutants induce far less IFN (1%–17%) than does wt virus (LAI and JOKLIK 1973). There is no correlation between the amount of IFN produced and any specific viral function, including the ability to synthesize dsRNA. Rather, the amount of IFN synthesized in mutant-infected L929 cells is correlated best with infectious virus yield. Neither defective top component particles which lack genome dsRNA nor subviral core particles induce IFN (LAI and JOKLIK 1973; HENDERSON and JOKLIK 1978).

The effect of UV irradiation on the induction of IFN has been examined in an attempt to better understand the molecular basis of the induction of IFN by reovirus. The IFN-inducing capacity of reovirus stocks is generally more resistant to inactivation by UV treatment than is virus infectivity (LAI and JOKLIK 1973; LONG and BURKE 1971). Furthermore, the IFN-inducing capacity of human and avian reoviruses is enhanced under certain conditions of UV irradiation (ELLIS et al. 1983; HENDERSON and JOKLIK 1978; LAI and JOKLIK 1973; LONG and BURKE 1971). Curiously, the IFN-inducing capacity at 38.5°C of the tsC447 mutant of human serotype 3 reovirus is greatly enhanced by UV irradiation. IFN induction by the UV-irradiated tsC mutant reovirions appears to involve, at least in part, the release of genome dsRNA molecules from the irradiated virion particles. HENDERSON and JOKLIK (1978) found that UV-irradiated tsC447 virions are not as stable as wt virions at 38.5°C, and concluded that the enhanced IFN-inducing capacity of the tsC mutant at 38.5°C likely is due to the release of the dsRNA genome segments from unstable particles. Indeed, dsRNA is a well-established inducer of IFN in mouse fibroblasts in culture (DEMAEYER and DEMAEYER-GUIGNARD 1988; HISCOTT et al. 1995; STEWART 1979).

Differences in IFN-inducing capacity between reovirus strains may be an important parameter contributing to the host response to reovirus infection (FIELDS and GREENE 1982). For example, reovirus-induced acute myocarditis in mice is correlated with viral RNA synthesis rather than with the generation of infectious virus in cardiac myocytes; it has been speculated that the IFN-inducing capacity of reoviruses may determine in part their potential for causing virus-induced acute myocarditis (SHERRY et al. 1996). Comparison of the ability of human reovirus to induce IFN in chick cells with that of a synthetic poly(rI)-poly(rC) dsRNA, which is an efficient IFN inducer, revealed that induction by the exogenously added synthetic dsRNA is considerably less sensitive to the effects of metabolic inhibitors (actinomycin and cycloheximide) than is induction by reovirus infection (LONG and BURKE 1971). Interestingly, a quantitative analysis of the dose (multiplicity)-response for production of IFN in aged chick embryo cells led WINSHIP and MARCUS (1980) to conclude that virion genome dsRNA constitutes the IFN inducer moiety of avian reovirus. In the permissive host cell the processing of genome dsRNA from most avian reovirus particles occurs naturally with a high probability, but for human reovirus this appears to be a much rarer event which may be intrinsic only to infectious virus and may require transcription for expression of the IFN-inducing capacity (WINSHIP and MARCUS 1980).

3 Antiviral Action of Interferon Against Reovirus

3.1 Sensitivity of Reovirus to the Antiviral Actions of Interferon

The effect of IFN treatment on the multiplication of reovirus is dependent upon the kind of host cell infected, the type of IFN, and the serotype of reovirus. Natural and molecularly cloned type I IFNs inhibit the multiplication of human reovirus as measured by the reduction in yield of infectious progeny virions in mouse ascites cells (GUPTA et al. 1974), mouse fibroblast cells (DEBENEDETTI et al. 1985a,b; SAMUEL and KNUTSON 1982a; WIEBE and JOKLIK 1975) and monkey kidney cells (DAHER and SAMUEL 1982; SAMUEL et al. 1982). Surprisingly, however, reovirus multiplication is not significantly inhibited by IFNs in all cell lines. For example, the multiplication of human reovirus is not significantly inhibited by either natural or cloned type I IFN-α in human amnion U (SAMUEL et al. 1982; SAMUEL and KNUTSON 1981), human fibroblast GM2767 (SAMUEL and KNUTSON 1981), human FS4 (RUBIN and GUPTA 1980), or most human HeLa cell lines (FEDUCHI et al. 1988; MUNOZ and CARRASCO 1984; NILSEN et al. 1982a), in marked contrast to the results obtained with mouse cell lines.

However, type I IFN-β does display an antiviral activity against reovirus in HeLa cells (NILSEN et al. 1982a). Also both natural and cloned type II IFN-γ inhibits reovirus multiplication in human fibroblast and human amnion cells (RUBIN and GUPTA 1980; SAMUEL and KNUTSON 1983). The extent to which reovirus replication is inhibited in mouse L cells by type I IFN-β depends also on the serotype of the challenge reovirus, with the serotype 3 Dearing strain more sensitive than the serotype 1 Lang strain (JACOBS and FERGUSON 1991). A saturating dose of

IFN-β, 1000U/ml, decreased Dearing strain replication 17- to 100-fold, whereas Lang strain was inhibited only two- to threefold.

3.2 Molecular Basis of the Antiviral Action of Interferon Against Reovirus

Human reovirus has been extensively studied in an attempt to elucidate the antiviral mechanism of IFN action. The earliest stages of the reovirus multiplication cycle at which IFN could conceivably act would be those involving the initiation of infection, that is, virion attachment to cells, penetration into cells, and uncoating within cells (Fig. 2). Studies in which the attachment of radioactively labeled reovirions and their uncoating to yield subviral particles were examined revealed that treatment of mouse cells with type I IFN does not significantly affect the attachment, penetration, or uncoating of parental reovirions under conditions where the production of infectious progeny is substantially reduced by IFN treatment (GALSTER and LENGYEL 1976; WIEBE and JOKLIK 1975).

The synthetic phase of the reovirus multiplication cycle corresponds to the time of viral macromolecular synthesis, that is, the production of viral mRNA transcripts, viral proteins and progeny genome dsRNA segments (JOKLIK 1983). This is the stage of the reovirus multiplication cycle most commonly inhibited in mouse and monkey cell lines by treatment with type I (α and β) IFN. The principal step of reovirus macromolecular synthesis inhibited by type I IFN in reovirus-infected mouse fibroblast and monkey kidney cells (DAHER and SAMUEL 1982; WIEBE and JOKLIK 1975) and reovirus vector-transfected monkey COS cells (GEORGE and SAMUEL 1988) is the translation of viral mRNA into viral protein.

The reovirus mutant tsC447 has been used to analyze the effect of IFN treatment on primary gene expression (SAMUEL et al. 1980; WIEBE and JOKLIK 1975). At the nonpermissive temperature of 39°C, essentially all of the viral mRNA synthesized in ts447-infected cells is produced by the virion-associated transcriptase from the parental subviral particles. When the accumulation of reovirus primary mRNA is quantified in IFN-treated as compared to untreated ts447-infected mouse L (WIEBE and JOKLIK 1975) or L929 (SAMUEL et al. 1980) fibroblasts, only a small IFN-dose dependent reduction in viral mRNA is observed. By contrast, the synthesis of reovirus proteins is drastically reduced in the IFN-treated cells infected either with the tsC447 mutant or with wild-type reovirus (GUPTA et al. 1974; SAMUEL et al. 1980; WIEBE and JOKLIK 1975). The reovirus late functions, whose occurrence is dependent upon the translation of primary or early transcripts, are all progressively more sensitive to IFN than is the translation of early mRNA (WIEBE and JOKLIK 1975).

The inhibitory effect of IFN-α and -β treatment on reovirus protein synthesis is selective. The synthesis of cellular proteins is not adversely affected by IFN treatment under conditions where reovirus protein synthesis is impaired (DAHER and SAMUEL 1982; DEBENEDETTI et al. 1985a; GUPTA et al. 1974; SAMUEL et al. 1980; WIEBE and JOKLIK 1975). Furthermore, the inhibitory effect of IFN treatment on reovirus protein synthesis observed in monkey cells is not observed for SV40 protein synthesis in cells doubly infected with SV40 and reovirus (DAHER and SAMUEL 1982). Reovirus protein synthesis is markedly inhibited in singly infected monkey CV1 and BSC cells that have been treated with type I IFN and then infected with reovirus. Likewise, reovirus protein synthesis is inhibited in cells doubly infected with reovirus and SV40 even though SV40 early protein synthesis is not inhibited, when IFN treatment is initiated after SV40 infection but before superinfection with reovirus (DAHER and SAMUEL 1982; KINGSMAN et al. 1980). The inhibitory effects of IFN treatment on reovirus protein synthesis observed in virus-infected cells are also observed in vector-transfected cells. When the reovirus S3 cDNA is inserted into an SV40-based vector, the encoded reovirus σNS protein is efficiently expressed in transfected monkey COS cells. However, the expression of the S3 gene is inhibited by type I IFN but not by type II IFN at the level of σNS protein synthesis (GEORGE and SAMUEL 1988). Selectivity of the IFN-induced inhibition of viral protein synthesis is also observed in HeLa cells, but in a different manner. IFN-α does not inhibit reovirus Dearing strain protein synthesis in a subclone of HeLa cells which are sensitive to the antiviral action of IFN, as demonstrated by superinfection with other RNA viruses (FEDUCHI et al. 1988). Superinfection of IFN-treated reovirus-infected HeLa cells with either poliovirus or vesicular stomatitis virus leads to an inhibition of synthesis of PV or VSV proteins, but not reovirus proteins (FEDUCHI et al. 1988).

The biochemical mechanism responsible for the inhibition of reovirus protein synthesis in IFN-sensitive mouse or monkey cells may involve multiple components of the protein synthesizing machinery that are functionally altered in IFN-treated, reovirus-infected cells (DEBENEDETTI et al. 1985a,b; DESROSIERS and LENGYEL 1979; FEDUCHI et al. 1988; NILSEN et al. 1982a,b; SAMUEL 1979; SAMUEL et al. 1984). For example, IFN treatment has been reported to inhibit reovirus mRNA 5′-cap methylation (DESROSIERS and LENGYEL 1979), to increase the degradation of reovirus mRNA in a selective manner (LENGYEL et al. 1980; SAMUEL et al. 1980), and to alter the functional activity of protein synthesis initiation factor eIF-2 in a selective manner (DEBENEDETTI et al. 1985a,b; NILSEN et al. 1982a; SAMUEL 1979; SAMUEL et al. 1984). Impairment of the initiation of reovirus mRNA translation is implicated as the principal cause of the IFN-mediated inhibition of reovirus protein synthesis, and alteration of protein synthesis initiation factor eIF-2α activity by phosphorylation appears to play a central role in this process (SAMUEL 1993).

The RNA-dependent protein kinase, PKR, is a key component in the control of protein synthesis in IFN-treated and virus-infected cells (CLEMENS 1996; SAMUEL 1991). PKR catalyzes the phosphorylation of eIF-2 on serine residue 51 of the alpha subunit (PATHAK et al. 1988; SAMUEL 1979). This modification of eIF-2α by phosphorylation leads to an inhibition of protein synthesis, because phosphorylation of eIF-2α blocks the eIF-2B catalyzed guanine nucleotide exchange reaction of eIF-2:GDP (CLEMENS 1996; SAMUEL 1993). IFN-α, IFN-β, and IFN-γ all induce PKR in mouse cells (SAMUEL and KNUTSON 1982a; SAMUEL 1986); in human cells, IFN-α is an efficient inducer but IFN-γ is a poor inducer of PKR (MEURS et al. 1990; THOMIS et al. 1992). The IFN-inducible cAMP-independent PKR acquires serine/threonine kinase activity following RNA-dependent autophosphorylation that at least in part occurs by an intermolecular mechanism (ATWATER and SAMUEL 1982; CLEMENS 1996; SAMUEL 1979, 1993; THOMIS and SAMUEL 1995). dsRNA, including reovirus genome dsRNA, is an efficient activator of PKR (SAMUEL 1979, 1993).

Phosphorylation of PKR, which reflects the autoactivation of PKR (SAMUEL 1993), is increased in IFN-treated cells infected with reovirus (GUPTA et al. 1982; SAMUEL et al. 1984). Analysis by two-dimensional isoelectric focusing/sodium dodecyl sulfate polyacrylamide gel electrophoresis and immunoblotting reveals that the phosphorylation of eIF-2α is increased in IFN-treated cells following infection with reovirus (SAMUEL et al. 1984). About 5%–10% of the eIF-2α is phosphorylated in untreated, uninfected cells, whereas 25%–30% is phosphorylated in IFN-treated mouse cells following infection with reovirus (SAMUEL et al. 1984). Likewise, the inhibition of reovirus protein synthesis that is observed following infection of one HeLa cell line with reovirus, but not another HeLa line, is correlated with an increase in phosphorylation of eIF-2α following IFN-β treatment (NILSEN et al. 1982a). In mouse fibroblasts, the kinetics of induction (SAMUEL and KNUTSON 1982a) and decay (SAMUEL and KNUTSON 1982b) of the IFN-induced PKR was correlated with the induction and decay of the antiviral state against reovirus.

The amount of reovirus mRNA that is polysome-associated is reduced in IFN-treated, as compared to untreated, infected L cells (DEBENEDETTI et al. 1985b). The binding of reovirus mRNA to ribosomes appears to be inhibited in IFN-treated L cells by a discriminatory mechanism that does not affect the translation of cellular mRNA (DEBENEDETTI et al. 1985b). Ribosome salt-wash fractions prepared from mouse ascites (SAMUEL and JOKLIK 1974) and mouse fibroblast (SAMUEL et al. 1977) treated with IFN possess an inhibitor of translation that impairs the translation of methylated, capped reovirus mRNA but not synthetic RNAs such as poly U. This ribosome-associated inhibitor is not species specific even though some IFNs are relatively species specific in their actions (SAMUEL and FARRIS 1977).

The IFN-induced PKR that mediates translation inhibition (SAMUEL 1993; CLEMENS 1996) has been purified and characterized from ribosome salt-wash fractions of mouse (BERRY et al. 1985) and human (SAMUEL et al. 1986) cell lines. PKR is now known to be one of three eukaryotic eIF-2α protein kinases (PKR, HRI, GCN2) that mediate translational control in response to various types of cell stress, including virus infection (SAMUEL 1993; CLEMENS 1996). Enhanced phosphorylation of PKR and eIF-2α is observed in the absence of detectable IFN-dependent mRNA degradation under conditions where reovirus mRNA translation is inhibited in mouse L cell-free systems (MIYAMOTO and SAMUEL 1980). The IFN-induced inhibition of viral mRNA translation in cell-free systems prepared from L cells can be reversed by addition of purified preparations of eIF-2 to the protein synthesizing reaction mixture (KAEMPFER et al. 1979).

Considerable progress has been made in our understanding of the structure, regulation and function of the PKR. PKR cDNA clones have been isolated and characterized (SAMUEL 1993). The human protein is 551 amino acids (THOMIS et al. 1992; KUHEN et al. 1996b) and the mouse protein is 515 amino acids (ICELY et al.

1991; TANAKA and SAMUEL 1995) as deduced from cDNA sequences. The kinase catalytic subdomains are located in the C-terminal region of PKR, and the RNA-binding subdomains are located in the N-terminal half of PKR (THOMIS et al. 1992; McCORMACK et al. 1992; SAMUEL 1993). PKR genomic clones have also been characterized (KUHEN et al. 1996a; TANAKA and SAMUEL 1994). The human *PKR* gene consists of 17 exons and spans about 50-kb on chromosome 2p21-22 (KUHEN et al. 1996a). The mouse *pkr* gene consists of 16 exons and spans about 28-kb on chromosome 17E2 (TANAKA and SAMUEL 1994; BARBER et al. 1993). $Pkr^{o/o}$ mice have been generated, and they show a reduced antiviral response induced by IFN-γ (YANG et al. 1995).

The major importance of PKR in the antiviral actions of IFN is illustrated by the fact that several viruses, including reovirus, have devised various strategies to antagonize the function of PKR. These strategies include: (a) the synthesis of viral dsRNA binding proteins such as the reovirus σ3 and vaccinia virus E3L proteins that sequester activator RNAs and thus antagonize the RNA-dependent autophosphorylation and activation of PKR (CHANG et al. 1992; MILLER and SAMUEL 1992; IMANI and JACOBS 1988); (b) the synthesis of viral proteins such as vaccinia virus K3L that mimic the eIF-2α substrate (BEATTIE et al. 1991; GALE et al. 1996); and (c) the synthesis of viral RNAs such as adenovirus VA RNA and Epstein-Barr virus EBER RNA that inhibit PKR activation by binding to PKR at the same site within the N-terminal region of the kinase that activator RNAs such as reovirus s1 mRNA bind (BISCHOFF and SAMUEL 1989; MATHEWS and SHENK 1991; McCORMACK et al. 1992; SAMUEL 1991). Heterodimerization of PKR with various cellular proteins, including the p58 inhibitor of PKR (GALE et al. 1996) and the HIV TAR RNA binding protein (BENKIRANE et al. 1997; COSENTINO et al. 1995), provides an additional mechanism for regulation of PKR activity.

The reovirus σ3 protein, which binds both synthetic and natural dsRNA molecules (HUISMANS and JOKLIK 1976; SCHIFF et al. 1988; MILLER and SAMUEL 1992) is a major outer capsid protein of reovirions (SMITH et al. 1969). The ability of cell-free extracts prepared from reovirus-infected cells to block the activation of PKR in vitro is due in large part to σ3 protein, and its ability to sequester dsRNA activators of PKR (IMANI and JACOBS 1988). Reovirus σ3 protein affects both translational efficiency (GIANTINI and SHATKIN 1989; SCHMECHEL et al. 1997) and IFN sensitivity (BEATTIE et al. 1995). Cotransfection of COS cells with S4 expression vectors enhances the expression of a CAT reporter, largely at the level of translation (GIANTINI and SHATKIN 1989; SELIGER et al. 1992). Presumably the functional form of the viral σ3 protein that enhances translation is not σ3 complexed with other viral proteins, because coexpression of S4 with the M2 cDNA that encodes the μ1/μ1c proteins does not mediate an enhanced expression of a CAT reporter whereas S4 alone does (TILLOTSON and SHATKIN 1992). The σ3 and μ1c proteins are known to form a complex (LEE et al. 1981).

Because the sequences of the S4 encoded σ3 proteins are highly conserved between the three serotypes of reovirus (ATWATER et al. 1986; KEDL et al. 1995; SELIGER et al. 1992), specifically including the region of the σ3 protein that possesses the dsRNA binding subdomain (MILLER and SAMUEL 1992), it seems unlikely that the dsRNA binding activity differs significantly among the σ3 proteins from serotypes 1, 2, and 3. Conceivably different amounts of free σ3 protein exist within cells infected with the different reovirus serotypes because of serotype-dependent differential complex formation between σ3 and μ1/μ1c (YUE and SHATKIN 1997). Thus, the relative sensitivity of the three serotypes to the antiviral action of IFN may possibly reflect a differential degree of antagonism of the PKR caused by a temporal quantitative difference in the amount of σ3 protein present in the state capable of affecting the balance of eIF-2α phosphorylation.

Although the mechanism of the translational enhancement attributed to the σ3 protein is generally assumed to derive from the dsRNA binding activity of σ3 that mediates a reduction in the available concentration of PKR activator RNAs, another intriguing possibility involves an alteration in the dephosphorylation of eIF-2αP. A curious sequence homology exists between the C-terminal region of the σ3 protein and a region of the regulatory subunit of the PP2A phosphatase, the enzyme responsible for the dephosphorylation of eIF-2αP (MILLER and SAMUEL 1992).

The fact that the reovirus σ3 protein affects IFN sensitivity is perhaps best illustrated by studies of cells infected with either vaccinia virus or adenovirus mutants, in which the reovirus S4 gene is used to complement the heterologous virus mutants defective in their ability to antagonize PKR function. For example, adenovirus that lacks the VAI gene is more sensitive to IFN and produces less late viral proteins than wild-type adenovirus because the activation of the PKR is not antagonized in cells infected with the VAI deletion mutant (KITAJEWSKI et al. 1986; MATHEWS and SHENK 1991). The adenovirus VAI RNA antagonizes PKR activation (SCHNEIDER et al. 1985) by binding to PKR at the same site as activator RNAs (MCCORMACK et al. 1992; MCCORMACK and SAMUEL 1995). The reovirus S4 gene encoding the σ3 protein will complement the VA deletion (LLOYD and SHATKIN 1992). Expression of a Ser51Ala substitution mutant of eIF-2α, which lacks the serine 51 phosphorylation site targeted by PKR, also complements the deletion of VAI (DAVIES et al. 1989). In the case of vaccinia virus, the E3L gene encodes a dsRNA binding protein (CHANG and JACOBS 1993). Vaccinia virus with the E3L gene deleted is more sensitive to IFN that is wild-type vaccinia in mouse cells. Insertion of the reovirus S4 gene, which encodes the σ3 dsRNA binding protein, into the E3L deletion mutant of vaccinia virus rescues E3L-deficient virus from the inhibitory actions of IFN (BEATTIE et al. 1995).

The 5′ cap structure of reovirus mRNA has been examined in IFN-treated cells. The amount of reovirus mRNA that is methylated at the 5′-terminal and penultimate G residues is not reduced by IFN treatment under conditions where virus yields are reduced by about 2 logs. However, reovirus mRNA that is 2′-O-methylated at the subsequent cytidine residue (cap 2 structure) is about 30%–50% lower for reovirus mRNA from IFN-treated than for untreated mouse L929 cells (DESROSIERS and LENGYEL 1979). The significance of this change remains unclear. Neither eIF-4A nor eIF-4B, two protein synthesis initiation factors that interact with mRNA at the 5′-cap during the translation initiation process (HERSHEY 1991), are altered in a detectable manner in IFN-treated or reovirus-infected mouse cells

(SAMUEL et al. 1984). Likewise, protein synthesis initiation factors eIF-4A, eIF-4B, eIF-3 and eIF-5 are not substrates for the purified PKR; purified PKR is highly selective for the α subunit of eIF-2 (BERRY et al. 1985). When RNA extracted from wild-type reovirus infected HeLa cells was examined by Northern blot analysis, full-sized reovirus mRNAs were detected in untreated but not in IFN-treated cells (NILSEN et al. 1982b). When primary transcripts specifically are analyzed, a small decrease in accumulation is observed in IFN-treated mouse cells (SAMUEL et al. 1980; WIEBE and JOKLIK 1975) but not in IFN-treated monkey CV1 cells which curiously possess slightly increased transcript levels (SAMUEL et al. 1980).

The efficiency of expression of the reovirus proteins encoded by the S1 and S4 genome segments differs at the level of protein synthesis (JOKLIK 1983). The IFN-inducible translational regulator PKR is implicated in controlling the differential translation of the s1 and s4 mRNAs (ATWATER et al. 1987; HENRY et al. 1994). PKR is typically induced about fivefold by IFN (KUHEN and SAMUEL 1997; SAMUEL 1993; THOMIS et al. 1992), and these elevated levels of the enzyme are believed to play an important role in the antiviral actions of IFN (SAMUEL 1991, 1993; CLEMENS 1996). However, the basal PKR present in the absence of treatment with exogenous IFN also plays an important role in the regulation of viral protein synthesis. This is illustrated, for example, by adenovirus (KITAJEWSKI et al. 1986; MATHEWS and SHENK 1991) and human immunodeficiency virus (BENKIRANE et al. 1997) as well as reovirus.

In the case of reovirus the S4 gene specifies a 1196 nt mRNA transcript that does not activate PKR; s4 mRNA is efficiently translated both in virus-infected and S4 vector-transfected cells. By contrast, the reovirus S1 gene specifies a 1463 nt mRNA transcript that is a potent activator of PKR autophosphorylation; s1 mRNA is inefficiently translated both in virus-infected and vector-transfected cells (ATWATER et al. 1987; BISCHOFF and SAMUEL 1989; GAILLARD and JOKLIK 1985; LEVIN and SAMUEL 1980; MUNEMITSU and SAMUEL 1988; RONER et al. 1989). Chimeric S1/S4 and S4/S1/S4 reovirus constructions that include the PKR activator sequence from S1 are expressed inefficiently, as S1 (HENRY et al. 1994). Treatment of COS cells with 2-aminopurine, an inhibitor of PKR, increases the translation of reovirus s1, s1/s4, and s4/s1/s4 chimeric transcripts that possess the PKR activator sequence, but not the translation of the s4 transcript that lacks the activator sequence (HENRY et al. 1994; SAMUEL and BRODY 1990). Likewise, co-expression of the phosphotransfer-negative mutant of PKR (K296R), the RNA-binding deficient PKR double mutant (K64E/K296R), or the truncated PKR (1–243) protein all increased the expression of S1 but not S4 in transfected cells (HENRY et al. 1994; ORTEGA et al. 1996; THOMIS and SAMUEL 1995).

The difference in translational efficiency observed between the reovirus s1 and s4 mRNAs attributed to PKR likely arises because of multiple reasons: the s1 mRNA includes a region that functions as an activator of PKR (BISCHOFF and SAMUEL 1989) which thus would lead to translational suppression; conversely, the s4 mRNA encodes the dsRNA binding protein σ3 (IMANI and JACOBS 1988; MILLER and SAMUEL 1992) that antagonizes PKR activation which thus would lead to translational stimulation (ATWATER et al. 1986; HENRY et al. 1994; SAMUEL and

BRODY 1990). The dsRNA binding protein E3L encoded by vaccinia virus is a potent inhibitor of PKR and is involved in the IFN-resistant phenotype of vaccinia virus (BEATTIE et al. 1995). Coexpression of the E3L protein increased S1 expression, but not S4 expression, in transfected COS cells (HENRY et al. 1994). These results suggest that the translational suppression in vivo mediated by PKR can be selective for mRNAs such as reovirus s1 mRNA that possess an intrinsic structure capable of activation of PKR.

dsRNA is an important effector of the cellular responses to IFN treatment following virus infection (JACOBS and LANGLAND 1996; PESTKA et al. 1987; VILCEK and SEN 1996). In addition to PKR, two other IFN-inducible enzymes have been identified that are also dsRNA-binding proteins: the 2′-5′-oligoadenylate synthetase (WILLIAMS and SILVERMAN 1985; SAMUEL 1991) and the dsRNA-specific adenosine deaminase (CATTANEO 1994; LIU et al. 1997; PATTERSON and SAMUEL 1995).

The 2′,5′-oligoadenylate pathway includes the synthetase that catalyzes the formation of the novel 2′,5′-oligoadenylates and RNase L, the endoribonuclease that is activated by 2′,5′-oligoA (SEN and RANSOHOFF 1992; WILLIAMS and SILVERMAN 1985; ZHOU et al. 1993). Although reovirus is not inhibited by IFN in most HeLa cell lines when grown in monolayer culture (MUNOZ and CARRASCO 1984; FEDUCHI et al. 1988), when grown in suspension culture reovirus protein synthesis is inhibited by IFN (NILSEN et al. 1982a,b). An increase in 2′,5′-oligoA is observed in IFN-treated as compared to untreated HeLa cells following reovirus infection (NILSEN et al. 1982b), and the cleavage of nascent reovirus mRNA by localized activation of the 2′,5′-oligoA-dependent RNase L has been described (BAGLIONI et al. 1984). Although concentrations of 2′,5′-oligoA sufficient to activate RNase L are produced in IFN-treated, reovirus-infected HeLa cells and a large fraction of the cellular mRNA is degraded (NILSEN et al. 1982b), protein synthesis is not significantly inhibited (NILSEN et al. 1983). However, in extracts of IFN-treated L cells under conditions where the 2′-5′ oligoA mediated degradation of reovirus mRNA is blocked the translation of reovirus mRNA is still inhibited by low but not by high concentrations of dsRNA in a manner that is correlated with the activation of PKR and eIF-2α phosphorylation (MIYAMOTO and SAMUEL 1980; MIYAMOTO et al. 1983). In addition, in a series of mouse cell clones derived from LMTK-1D cells, greatly increased levels of 2′,5′-oligoA synthetase can occur without activation of an antiviral state; conversely, an antiviral state is exhibited without a detectable increase in 2′,5′-oligoA synthetase activity (LEWIS 1988). Although the 2′,5′-oligoA synthetase-RNase L pathway does not appear to play a central role in the antiviral action of IFN against reovirus, the pathway is an important component of the IFN response in the case of other viruses. For example, expression of cDNA clones encoding the 2′,5′-oligoA synthetase in hamster and mouse cells is sufficient to establish an antiviral state against picornaviruses, but not other viruses (CHEBATH et al. 1987; COCCIA et al. 1990).

The third IFN-inducible, dsRNA binding enzyme is an adenosine deaminase specific for RNA (ADAR). ADAR catalyzes the covalent modification of dsRNA substrates by hydrolytic C-6 deamination of adenosine to yield inosine (BASS and WEINTRAUB 1988; CATTANEO 1994). Two forms of ADAR are present in human cells: an IFN-inducible 150-kDa protein present in both the cytoplasm and nucleus, and a 110-kDa protein present predominantly if not exclusively in the nucleus (Patterson et al. 1995; Patterson and Samuel 1995). The ADAR deaminase possesses three copies of the highly conserved dsRNA-binding motif (Patterson and Samuel 1995) originally discovered in the PKR which possesses two copies of the motif (McCormack et al. 1992). The three dsRNA-binding domains present in the ADAR deaminase are functionally distinct (Liu and Samuel 1996; Liu et al. 1997).

ADAR is implicated in two types of RNA editing processes. First, A–I modifications are found at multiple sites in viral RNAs, as exemplified by the biased hypermutations observed in negative-stranded RNA virus genomes during lytic and persistent infections; second, highly selective C-6 adenosine deaminations occur at one or a few sites in certain viral and cellular RNAs as exemplified by hepatitis D virus RNA and the GluR receptor channel pre-mRNAs (Cattaneo 1994; Liu et al. 1997). The role, if any, that the dsRNA-specific ADAR editing enzyme plays in the replication of reovirus, in untreated or IFN-treated cells, remains to be established.

Finally, the expression of both class I and class II major histocompatibility antigens (MHC) is increased in IFN-treated cells, class I MHC by IFN-α, -β, and -γ, and class II MHC by IFN-γ (Samuel 1991). MHC antigens are essential for immunocompetent cells to present many foreign antigens, including peptides derived from virus proteins, to T cells in the generation of specific immune responses (Townsend and Bodmer 1989; Yewdell and Bennink 1990). The IFN-induced expression of MHC may potentially contribute to the antiviral actions of IFN against reoviruses within the whole animal by enhancement of cellular immune responses.

4 Summary

Reovirus induces IFN, and reovirus is sensitive to the antiviral actions of IFN. The characteristics of the IFN-inducing capacity of reovirus, and the antiviral actions of IFN exerted against reovirus, are dependent upon the specific combination of reovirus strain, host cell line, and IFN type. Responses, both IFN induction and IFN action, differ quantitatively if not qualitatively and are dependent upon the virus, cell, and IFN combination. Stable natural dsRNA, identified as the form of nucleic acid that constitutes the reovirus genome, is centrally involved in the function of at least three IFN-induced enzymes. Protein phosphorylation by PKR, RNA editing by the ADAR adenosine deaminase, and RNA degradation by the 2′,5′-oligoA pathway all involve dsRNA either as an effector or as a substrate.

Considerable evidence implicates PKR as a particularly important contributor to the IFN-induced antiviral state displayed at the level of the single virus-infected cell, where the translation of viral mRNA is often observed to be inhibited following treatment with IFN-α/β. In the whole animal infected with reovirus, elevated cellular immune responses mediated by enhanced expression of MHC class I and class II antigens induced by IFN-α/β or IFN-γ may contribute significantly to the overall antiviral response.

References

Atta MS, Irving Wl, Powell RJ, Todd I (1995) Enhanced expression of MHC class I molecules on cultured human thyroid follicular cells infected with reovirus through induction of type 1 interferons. Clin Exp Immunol 101:121–126

Atwater JA, Samuel CE (1982) Interferon inhibits the synthesis of viral proteins and induces P1 phosphorylation in both adenylate cyclase-deficient and cAMP-dependent protein kinase deficient variants of mouse lymphoma cells. Virology 123:206–211

Atwater JA, Munemitsu SM, Samuel CE (1986) Molecular cDNA cloning and nucleotide sequence of the reovirus serotype 1 Lang strain S4 mRNA which encodes the major capsid surface polypeptide σ3. Biochem Biophys Res Commun 136:183–192

Atwater JA, Munemitsu SM, Samuel CE (1987) Efficiency of expression of cDNAs of the reovirus S1 and S4 genes in transfected animal cells differs at the level of translation. Virology 159:350–357

Baglioni C, De Benedetti A, Williams GJ (1984) Cleavage of nascent reovirus mRNA by localized activation of the 2′-5′-oligoadenylate-dependent endoribonuclease. J Virol 52:865–871

Barber GN, Edelhoff S, Katze MG, Disteche CM (1993) Chromosomal assignment of the interferon-inducible double-stranded RNA-dependent protein kinase (PRKR) to human chromosome 2p21-p22 and mouse chromosome 17E2. Genomics 16:765–767

Bass BL, Weintraub H (1988) An unwinding activity that covalently modified its double-stranded RNA substrate. Cell 55:1089–1098

Beattie E, Tartaglia J, Paoletti E (1991) Vaccinia virus-encoded eIF-2α homologue abrogates the antiviral effect of interferon. Virology 183:419–422

Beattie E, Denzler KL, Tartaglia J, Perkus ME, Paoletti E, Jacobs BL (1995) Reversal of the interferon-sensitive phenotype of a vaccina virus lacking E3L by expression of the reovirus S4 gene. J Virol 69:499–505

Benkirane M, Neuveut C, Chun RF, Smith SM, Samuel CE, Gatignol A, Jeang KT (1997) Oncogenic potential of TAR RNA-binding protein TRBP and its regulatory interaction with RNA-dependent protein kinase PKR. EMBO J 16:611–624

Berry MJ, Knutson GS, Lasky SR, Munemitsu SM, Samuel CE (1985) Mechanism of interferon action. Purification and substrate specificities of the double-stranded RNA-dependent protein kinase from untreated and interferon-treated mouse fibroblasts. J Biol Chem 260:11240–11247

Bischoff JR, Samuel CE (1989) Activation of the P1/eIF-2a protein kinase by individual reovirus s-class mRNAs: s1 mRNA is a potent activator relative to s4 mRNA. Virology 172:106–115

Blatt LM, Davis JM, Klein SB, Taylor MW (1996) The biologic activity and molecular characterization of a novel synthetic interferon-alpha species, consensus interferon. J Interferon Cytokine Res 16: 489–499

Campbell IL, Harrison LC (1989) Viruses and cytokines: evidence for multiple roles in pancreatic beta cell destruction in type 1 insulin-dependent diabetes mellitus. J Cell Biochem 40:57–66

Cattaneo R (1994) Biased (A > I) hypermutation of animal RNA virus genomes. Curr Opin Genet Dev 4:895–900

Chang HW, Jacobs BL (1993) Identification of a conserved motif that is necessary for binding of the vaccinia virus E3L gene products to double-stranded RNA. Virology 194:537–547

Chang HW, Watson JC, Jacobs BL (1992) The E3L gene of vaccinia virus encodes an inhibitor of the interferon-induced double-stranded RNA-dependent protein kinase. Proc Natl Acad Sci USA 89:4825–4829

Chebath J, Benech P, Revel M, Vigneron M (1987) Constitutive expression of 2′5′ oligo A synthetase confers resistance to picornavirus infection. Nature 330:587–588

Clemens MJ (1996) Protein kinases that phosphorylate eIF-2 and eIF-2B, and their role in eukaryotic cell translational control. In: Hershey JWB, Mathews MB, Sonenberg N (eds) Translational control. Cold Spring Harbor Press, Cold Spring Harbor, NY, pp 139–172

Coccia EM, Romeo G, Nissim A, Marziali G, Albertini R, Affabris E, Battistini A, Fiorucci G, Orsatti R, Rossi GB, Chebath J (1990) A full-length murine 2–5A synthetase cDNA transfected in NIH-3T3 cells impairs EMCV but not VSV replication. Virology 179:228–233

Cosentino GP, Venkatesan S, Serluca FC, Green SR, Mathews MB, Sonenberg N (1995) Double-stranded RNA-dependent protein kinase and TAR RNA-binding protein form homo- and heterodimers in vivo. Proc Natl Acad Sci USA 92:9445–9449

Daher KA, Samuel CE (1982) Mechanism of interferon action. Differential effect of interferon on the synthesis of simian virus 40 and reovirus polypeptides in monkey kidney cells. Virology 117:379–390

Darnell JE Jr, Kerr IM, Stark GM (1994) Jak-STAT pathways and transcriptional activation in response to IFNs and other extracellular signaling proteins. Science 264:1415–1421

Davies MV, Furtado M, Hershey JWB, Thimmappaya B, Kaufman RJ (1989) Complementation of adenovirus virus-associated RNA I gene deletion by expression of a mutant eukaryotic translation initiation factor. Proc Natl Acad Sci USA 86:9163–9167

DeBenedetti A, Williams GJ, Comeau L, Baglioni C (1985a) Inhibition of viral mRNA translation in interferon-treated L cells infected with reovirus. J Virol 55:588–593

DeBenedetti A, Williams GJ, Baglioni C (1985b) Inhibition of binding to initiation complexes of nascent reovirus mRNA by double-stranded RNA-dependent protein kinase. J Virol 54:408–413

DeMaeyer E, DeMaeyer-Guignard J (1988) Interferons and other regulatory cytokines. Wiley, New York

Desrosiers RC, Lengyel P (1979) Impairment of reovirus mRNA "cap" methylation in interferon-treated mouse L929 cells. Biochim Biphys Acta 562:471–480

Diaz MO, Bohlander S, Allen G (1993) Nomenclature of the human interferon genes. J Interferon Res 13:61–62

Ellis MN, Eidson CS, Brown J, Kleven SH (1983) Studies on interferon induction and interferon sensitivity of avian reoviruses. Avian Dis 27:927–936

Feduchi E, Esteban M, Carrasco L (1988) Reovirus type 3 synthesizes proteins in interferon-treated HeLa cells without reversing the antiviral state. Virology 164:420–426

Fields BN, Greene MI (1982) Genetic and molecular mechanisms of viral pathogenesis: implications for prevention and treatment. Nature 300:19–32

Galster RL, Lengyel P (1976) Formation and characteristics of reovirus subviral particles in interferon-treated mouse L cells. Nucleic Acid Res 3:581–598

Gaillard RK, Joklik WK (1985) The relative translation efficiency of reovirus messenger RNAs. Virology 147:336–348

Gale M, Tan SL, Wambach M, Katze MG (1996) Interaction of the interferon-induced PKR protein kinase with inhibitory proteins p58 and vaccinia virus K3L is mediated by unique domains: implication for kinase regulation. Mol Cell Biol 16:4172–4181

George CX, Samuel CE (1988) Mechanism of interferon action. Expression of reovirus S3 gene in transfected COS cells and subsequent inhibition at the level of protein synthesis by type I but not by type II interferon. Virology 166:573–582

Giantini M, Shatkin AJ (1989) Stimulation of chloramphenicol acetyltransferase mRNA translation by reovirus capsid polypeptide σ3 in cotransfected COS cells. J Virol 63:2415–2421

Gupta SL, Graziadi WD, Weideli H, Sopori M, Lengyel P (1974) Selective inhibition of viral protein accumulation in interferon-treated cells: nondiscriminate inhibition of the translation of added viral and cellular messenger RNAs in their extracts. Virology 57:49–63

Gupta SL, Holmes SL, Mehra LL (1982) Interferon action against reovirus: activation of interferon-induced protein kinase in mouse L929 cells upon reovirus infection. Virology 120:495–499

Henderson DR, Joklik WK (1978) The mechanism of interferon induction by UV-irradiated reovirus. Virology 91:389–406

Henry GL, McCormack SJ, Thomis DC, Samuel CE (1994) Translational control and the RNA-dependent protein kinase PKR: antagonists of PKR enhance the translational activity of mRNAs that include a 161 nucleotide region from reovirus s1 mRNA. J Biol Regul Homeost Agents 8:15–24

Hershey JWB (1991) Translational control in mammalian cells. Annu Rev Biochem 60:715–755

Hiscott J, Nguyen M, Lin RT (1995) Molecular mechanisms of interferon beta gene induction. Semin Virol 6:161–173

Huismans H, Joklik WK (1976) Reovirus-coded polypeptides in infected cells: isolation of two native monomeric polypeptides with affinity for single-stranded and double-stranded RNA respectively. Virology 70:411–424

Icely PL, Gross P, Bergeron JJM, Devault A, Afar DEH, Bell JC (1991) TIK, a novel serine/threonine kinase, is recognized by antibodies directed against phosphotyrosine. J Biol Chem 266: 16073–16077

Imani F, Jacobs BL (1988) Inhibitory activity for the interferon-induced protein kinase is associated with the reovirus serotype 1 sigma 3 protein. Proc Natl Acad Sci USA 85:7887–7891

Isaacs A, Lindenmann J (1957) Virus interference. I. The interferon. Proc R Soc Lond B 147:258–267

Jacobs BL, Ferguson RE (1991) The Lang strain of reovirus serotype 1 and the Dearing strain of reovirus serotype 3 differ in their sensitivities to beta interferon. J Virol 65:5102–5104

Jacobs BL, Langland JO (1996) When two strands are better than one: the mediators of modulators of cellular responses to double-stranded RNA. Virology 219:339–349

Joklik WK (1983) The reoviridae. Plenum, New York

Kaempfer R, Israeli R, Rosen H, Knoller S, Zilberstein A, Schmidt A, Revel M (1979) Reversal of the interferon-induced block of protein synthesis by purified preparation of eukaryotic initiation factor 2. Virology 99:170–173

Kedl R, Schmechel S, Schiff L (1995) Comparative sequence analysis of the reovirus S4 genes from 13 serotype 1 and serotype 3 field isolates. J Virol 69:552–559

Kingsman SM, Smith MD, Samuel CE (1980) Mechanism of interferon action. Simian virus 40-specific early polypeptides synthesized in untreated and interferon-treated monkey kidney cells. Proc Natl Acad Sci USA 77:2419–2423

Kitajewski J, Schneider RJ, Safer B, Munemitsu SM, Samuel CE, Thimmappaya B, Shenk T (1986) Adenovirus VAI RNA antagonizes the antiviral state of interferon by preventing activation of the interferon-induced eIF-2a kinase. Cell 45:195–200

Kuhen KL, Samuel CE (1997) Isolation of the interferon-inducible RNA-dependent protein kinase Pkr promoter and identification of a novel DNA element within the 5′-flanking region of the human and mouse Pkr genes. Virology 227:119–130

Kuhen KL, Shen X, Carlisle ER, Richardson AL, Weier HUG, Tanak H, Samuel CE (1996a) Structural organization of the human gene PKR encoding an interferon-inducible RNA-dependent protein kinase and differences from its mouse homolog. Genomics 36:197–201

Kuhen KL, Shen X, Samuel CE (1996b) Mechanism of interferon action. Sequence of the human interferon-inducible RNA-dependent protein kinase (PKR) deduced from genomic clones. Gene 178:191–193

Lai M-HT, Joklik WK (1973) The induction of interferon by temperature-sensitive mutants of reovirus, UV-irradiated reovirus, and subviral reovirus particles. Virology 51:191–204

Lee PWK, Hayes EC, Joklik WK (1981) Characterization of anti-reovirus immunoglobulins secreted by cloned hybridoma cell lines. Virology 108:134–146

Lengyel P, Desrosiers R, Broeze R, Slattery E, Taira H, Dougherty J, Samanta H, Pichon J, Farrell P, Ratner L, Sen G (1980) In microbiology 1980. American Society of Microbiology, Washington DC, pp 219–226

Levin KH, Samuel CE (1980) Biosynthesis of reovirus-specified polypeptides. Purification and characterization of the small-sized class mRNAs of reovirus type 3: coding assignments and translational efficiencies. Virology 106:1–13

Lewis JA (1988) Induction of an antiviral state by interferon in the absence of elevated levels of 2,5-oligo(A) synthetase and eIF-2 kinase. Virology 162:118–127

Liu Y, Samuel CE (1996) Mechanism of interferon action. Functionally distinct dsRNA-binding and catalytic domains in the interferon-inducible, double-stranded RNA-specific adenosine deaminase. J Virol 70:1961–1968

Liu Y, George CX, Patterson JB, Samuel CE (1997) Functionally distinct dsRNA-binding domains associated with alternative splice-site variants of the interferon-inducible double-stranded RNA-specific adenosine deaminase. J Biol Chem 272:4419–4428

Lloyd RM, Shatkin AJ (1992) Translational stimulation by reovirus polypeptide σ3: substitution for VAI RNA and inhibition of phosphorylation of the α subunit of eukaryotic initiation factor eIF-2. J Virol 66:6878–6884

Long WF, Bürke DC (1971) Interferon production by double-stranded RNA: a comparison of interferon induction by reovirus RNA to that by a synthetic double-stranded polynucleotide. J Gen Virol 12: 1–11

Major AS, Cuff CF (1996) Effects of the route of infection on immunoglobulin G subclasses and specificity of the reovirus-specific humoral immune response. J Virol 70:5968–5974

Mathews MB, Shenk T (1991) Adenovirus virus-associated RNA and translational control. J Virol 65:5657–5662

McCormack SJ, Samuel CE (1995) Mechanism of interferon action. RNA-binding activity of full-length and R-domain forms of the RNA-dependent protein kinase PKR – determination of KD values for VAI and TAR RNAs. Virology 206:511–519

McCormack SJ, Thomis DC, Samuel CE (1992) Mechanism of interferon action. Identification of a RNA-binding domain within the N-terminal region of the human RNA-dependent P1/eIF-2α protein kinase. Virology 188:47–56

Meurs E, Chong K, Galabru J, Thomas NSB, Kerr IM, Williams BRG, Hovanessian AG (1990) Molecular cloning and characterization of the human double-stranded RNA-activated protein kinase induced by interferon. Cell 62:379–390

Miller JE, Samuel CE (1992) Proteolytic cleavage of the reovirus sigma 3 protein results in enhanced double-stranded RNA-binding activity: identification of a repeated basic amino acid motif within the C-terminal binding region. J Virol 66:5347–5356

Miyamoto NG, Samuel CE (1980) Mechanism of interferon action. Interferon-mediated inhibition of reovirus mRNA translation in the absence of detectable mRNA degradation but in the presence of protein phosphorylation. Virology 107:461–475

Miyamoto NG, Jacobs BL, Samuel CE (1983) Mechanism of interferon action. Effect of double-stranded RNA and the 5′-O-monophosphate form of 2′,5′-oligoadenylate on the inhibition of reovirus mRNA translation in vitro. J Biol Chem 258:15232–15237

Munemitsu SM, Samuel CE (1988) Biosynthesis of reovirus-specified polypeptides: effect of point mutation of the sequences flanking the 5′-proximal AUG initiation codons of the reovirus s1 and s4 genes on the efficiency of mRNA translation. Virology 163:643–646

Munoz A, Carrasco L (1984) Action of human lymphoblastoid interferon on HeLa cells infected with RNA-containing animal viruses. J Gen Virol 65:377–390

Nagano Y, Kojima Y (1958) Inhibition de l'infection vaccinale par le virus homologue. CR Seances Soc Bio Filiales 152:1627–1630

Neufeld DS, Platzer M, Davies TF (1989) Reovirus induction of MHC class II antigen in rat thyroid cells. Endocrinology 124:543–545

Nilsen TW, Maroney PA, Baglioni C (1982a) Inhibition of protein synthesis in reovirus-infected HeLa cells with elevated levels of interferon-induced protein kinase activity. J Biol Chem 257:14593–14596

Nilsen TW, Maroney PA, Baglioni C (1982b) Synthesis of (2′-5′)oligoadenylate and activation of an endoribonuclease in interferon-treated HeLa cells infected with reovirus. J Virol 42:1039–1045

Nilsen TW, Maroney PA, Baglioni C (1983) Maintenance of protein synthesis in spite of mRNA breakdown in interferon-treated HeLa cells infected with reovirus. Mol Cell Biol 3:64–69

Ortega LG, McCotter MD, Henry GL, McCormack SJ, Thomis DC, Samuel CE (1996) Mechanism of interferon action. Biochemical and genetic evidence for the intermolecular association of the RNA-dependent protein kinase PKR from human cells. Virology 215:31–39

Pathak VK, Schindler D, Hershey JWB (1988) Generation of a mutant form of protein synthesis initiation factor eIF-2 lacking the site of phosphorylation by eIF-2 kinases. Mol Cell Biol 8:993–995

Patterson JB, Samuel CE (1995) Expression and regulation by interferon of a double-stranded RNA-specific adenosine deaminase from human cells: evidence for two forms of the deaminase. Mol Cell Biol 15:5376–5388

Patterson JB, Thomis DC, Hans SL, Samuel CE (1995) Mechanism of interferon action. Double-stranded RNA-specific adenosine deaminase from human cells is inducible by alpha and gamma interferons. Virology 210:508–511

Pestka S, Langer JA, Zoon KC, Samuel CE (1987) Interferons and their actions. Annu Rev Biochem 56:727–777

Roner MR, Gaillard RK, Joklik WK (1989) Control of reovirus mRNA translation efficiency by the regions upstream of initiation codons. Virology 168:292–301

Rubin BY, Gupta SL (1980) Differential efficacies of human type I and type II interferons as antiviral and antiproliferative agents. Proc Natl Acad Sci USA 77:5928–5932

Samuel CE (1979) Mechanism of interferon action. Phosphorylation of protein synthesis initiation factor eIF-2 in interferon-treated human cells by a ribosome-associated protein kinase possessing site-specificity similar to hemin-regulated rabbit reticulocyte kinase. Proc Natl Acad Sci USA 76:600–604

Samuel CE (1986) Relationship between the interferon-induced antiviral state and protein P1/eIF-2α phosphorylation in mouse fibroblasts treated with natural and recombinant viral and immune interferon. In: The biology of the interferon system 1985. Elsevier, Amsterdam, pp 101–110

Samuel CE (1991) Antiviral actions of interferon. Interferon-regulated cellular proteins and their surprisingly selective antiviral activities. Virology 183:1–11

Samuel CE (1993) The eIF-2α protein kinases, regulators of translation in eukaryotes from yeasts to humans. J Biol Chem 268:7603–7606

Samuel CE, Brody MS (1990) Biosynthesis of reovirus-specified polypeptides: 2-aminopurine increases the efficiency of translation of reovirus s1 mRNA but not s4 mRNA in transfected cells. Virology 176:106–113
Samuel CE, Farris DA (1977) Mechanism of interferon action. Species specificity of interferon and of the interferon-mediated inhibitor of translation from mouse, human and monkey cells. Virology 77: 556–565
Samuel CE, Joklik WK (1974) A protein synthesizing system from interferon treated cells that discriminates between cellular and viral messenger RNAs. Virology 74:403–413
Samuel CE, Knutson GS (1981) Mechanism of interferon action: cloned human leukocyte interferons induce protein kinase and inhibit vesicular stomatitis virus but not reovirus replication in human amnion cells. Virology 114:302–306
Samuel CE, Knutson GS (1982a) Mechanism of interferon action. Kinetics of induction of the antiviral state and protein phosphorylation in mouse fibroblasts treated with natural and cloned interferons. J Biol Chem 257:11791–11795
Samuel CE, Knutson GS (1982b) Mechanism of interferon action. Kinetics of decay of the antiviral state and protein phosphorylation in mouse fibroblasts treated with natural and cloned interferons. J Biol Chem 257:11796–11801
Samuel CE, Knutson GS (1983) Mechanism of interferon action: human leukocyte and immune interferons regulate the expression of different genes and induce different antiviral states in human amnion U cells. Virology 130:474–484
Samuel CE, Farris DA, Eppstein DA (1977) Mechanism of interferon action. Kinetics of interferon action in mouse L929 cells: translation inhibition, protein phosphorylation, and messenger RNA methylation and degradation. Virology 83:56–71
Samuel CE, Kingsman SM, Melamed RW, Farris D, Smith MD, Miyamoto NG, Lasky SR, Knutson GS (1980) Mechanism of interferon mediated inhibition of protein synthesis. Ann NY Acad Sci 350: 475–485
Samuel CE, Knutson GS, Masters PS (1982) Ability of cloned human type-alpha interferons to induce protein phosphorylation and inhibit virus replication is specified by the host cell rather than the interferon subspecies. J Interferon Res 2:563–574
Samuel CE, Duncan R, Knutson GS, Hershey JW (1984) Mechanism of interferon action. Increased phosphorylation of protein synthesis initiation factor eIF-2 alpha in interferon-treated, reovirus-infected mouse L929 fibroblasts in vitro and in vivo. J Biol Chem 259:13451–13457
Samuel CE, Knutson GS, Berry MJ, Atwater JA, Lasky SR (1986) Purification of double-stranded RNA-dependent protein kinase from mouse fibroblasts. Methods Enzymol 119:499–516
Schiff LA, Nibert ML, Co MS, Brown EG, Fields BN (1988) Distinct binding sites for zinc and double-stranded RNA in the reovirus outer capsid protein σ3. Mol Cell Biol 8:273–283
Schindler C, Darnell JE Jr (1995) Transcriptional responses to polypeptide ligands: the JAK-STAT pathway. Annu Rev Biochem 64:621–651
Schmechel S, Chute M, Skinner P, Anderson R, Schiff L (1997) Preferential translation of reovirus mRNA by a sigma3-dependent mechanism. Virology 232:62–73
Schneider RJ, Safer B, Munemitsu SM, Samuel CE, Shenk T (1985) Adenovirus VAI RNA prevents phosphorylation of the eukaryotic initiation factor 2 α subunit subsequent to infection. Proc Natl Acad Sci USA 82:4321–4325
Sen GC, Ransohoff RM (1992) Interferon-induced antiviral actions and their regulation. Adv Virus Res 42:57–102
Sharpe AH, Fields BN (1983) Pathogenesis of reovirus infection. In: Joklik WK (ed) The reoviridae. Plenum, New York, pp 229–285
Seliger LS, Giantini M, Shatkin AJ (1992) Translational effects and sequence comparisons of the three serotypes of the reovirus S4 gene. Virology 187:202–210
Sherry B, Bàty CJ, Blum MA (1996) Reovirus-induced acute myocarditis in mice correlates with viral RNA synthesis rather than generation of infectious virus in cardiac myocytes. J Virol 70:6709–6715
Smith RE, Zweerink HJ, Joklik WK (1969) Polypeptide components of virions, top component, and cores of reovirus type 3. Virology 39:791–798
Staeheli P (1990) Interferon-induced proteins and the antiviral state. Adv Virus Res 38:147–200
Stewart WE (1979) The interferon system. Springer, Berlin Heidelberg New York
Tanaka H, Samuel CE (1994) Mechanism of interferon action. Structure of the mouse PKR gene encoding the interferon-inducible RNA-dependent protein kinase. Proc Natl Acad Sci USA 91:7995–7999
Tanaka H, Samuel CE (1995) Sequence of the murine interferon-inducible RNA-dependent protein kinase PKR deduced from genomic clones. Gene 153:283–284

Thomis DC, Samuel CE (1995) Characterization of the intermolecular autophosphorylation of PKR, the interferon-inducible RNA-dependent protein kinase. J Virol 69:5195–5198
Thomis DC, Doohan JP, Samuel CE (1992) Mechanism of interferon action: cDNA structure, expression and regulation of the interferon-induced, RNA-dependent Pl/eIF-2a protein kinase from human cells. Virology 188:33–46
Tillotson L, Shatkin AJ (1992) Reovirus polypeptide σ3 and N-terminal myristoylation of polypeptide μ1 are required for site-specific cleavage to μ1C in transfected cells. J Virol 66:2180–2186
Townsend A, Bodmer H (1989) Antigen recognition by class I restricted T lymphocytes. Annu Rev Immunol 7:601–624
Vilcek J, Sen GC (1996) Interferons and other cytokines. In: Fields BN, Knipe DM, Howley PM (eds) Fundamental virology. Lippincott-Raven, Philadelphia, pp 341–365
Wiebe ME, Joklik WK (1975) The mechanism of inhibition of reovirus replication by interferon. Virology 66:229–240
Williams BRG, Silverman RH (1985) The 2–5A system: molecular and clinical aspects of the interferon-regulated pathway. Prog Clin Biol Res 202:1–478
Winship TR, Marcus PI (1980) Interferon induction by viruses. VI. Reovirus: virion genome dsRNA as the interferon inducer in aged chick embryo cells. J Interferon Res 1:155–167
Yang YL, Reis LFL, Pavlovic J, Aguzzi A, Schafer R, Kumar A, Williams BRG, Aguet M, Weissmann C (1995) Deficient signaling in mice devoid of double-stranded RNA-dependent protein kinase. EMBO J 14:6095–6106
Yue Z, Shatkin AJ (1997) Double-stranded RNA-dependent protein kinase (PKR) is regulated by reovirus structural proteins. Virology 234:364–371
Yewdell JW, Bennink JR (1990) The binary logic of antigen processing and presentation to T cells. Cell 62:203–206
Young HA (1996) Regulation of interferon-γ gene expression. J Interferon Cytokine Res 16:563–568
Zhou A, Hassel BA, Silverman RH (1993) Expression cloning of 2–5A-dependent RNAase: a uniquely regulated mediator of interferon action. Cell 72:753–765

Cellular and Humoral Immunity to Reovirus Infection

H.W. Virgin[1], T.S. Dermody[2] and K.L. Tyler[3,4]

1 Introduction

Mammalian reoviruses are nonenveloped double-stranded RNA (dsRNA) viruses of the genus *Orthoreovirus*, family Reoviridae. The biology of reoviruses has been

[1]Center for Immunology and Division of Infectious Diseases, Departments of Pathology and Molecular Microbiology, Washington University School of Medicine, Box 8051, 660 South Euclid Ave., St. Louis, MO 63110, USA E-mail: virgin@immunology.wustl.edu
[2]Departments of Pediatrics, and Microbiology and Immunology and Elizabeth B. Lamb Center for Pediatric Research, Vanderbilt University School of Medicine, Nashville, TN 37232, USA
E-mail: terry.dermody@mcmail.vanderbilt.edu
[3]Departments of Neurology, and Microbiology, Immunology, and Medicine, University of Colorado Health Sciences Center, 4200 East 9th Ave., Denver, CO 80262, USA E-mail: ken.tyler@uchsc.edu
[4]Neurology Service, Denver Veterans Affairs Medical Center, 1055 Clermont St., Denver, CO 80220, USA

recently reviewed in detail (VIRGIN et al. 1997; NIBERT et al. 1995; TYLER and FIELDS 1995). This chapter focuses on studies of immunity to reovirus.

Reoviruses infect humans but are not associated with significant morbidity or mortality (reviewed in VIRGIN et al. 1997; TYLER and FIELDS 1995). Despite the lack of a specific human disease association, experimental infection of animals with reovirus has been extremely valuable for analyzing how viruses cause disease (pathogenesis), the mechanisms underlying the severity of disease (virulence), and the basis for associations between infection and damage to specific cells or tissues (tropism). Indeed, principles that govern our current understanding of viral pathogenesis frequently have been derived from studies of reovirus pathogenesis (reviewed in VIRGIN et al. 1997; MORRISON and FIELDS 1991; NIBERT et al. 1991). The primary experimental models for analysis of reovirus pathogenesis has been infection of newborn or, more recently, adult severe combined immunodeficient (SCID) mice (HALLER et al. 1995a,b; GEORGE et al. 1990; SHERRY et al. 1993).

The application of genetic techniques to analysis of viral pathogenesis has been a particularly important aspect of studies in the reovirus model system. Due to the segmented nature of the reovirus genome, genetic approaches allow identification of genome segments, and therefore proteins, involved in a particular stage of pathogenesis or step in infection of the cell. More recently structural information about the virion and structural intermediates in the virus replication cycle has clarified interpretation of both pathogenetic and genetic studies. The extensive background knowledge of stages of reovirus pathogenesis in animals and steps in reovirus infection of cells has facilitated analysis of how the immune system, especially humoral immunity, protects the host. This chapter reviews the immune response to reovirus with a special emphasis on how structural, genetic, and cell biological information can contribute to understanding mechanisms of immunity.

2 Host Factors Contributing to Viral Tropism

2.1 Role of Host and Immune Factors in Patterns of Viral Disease

Reoviruses have been particularly useful for defining viral genes involved at different stages of pathogenesis. While much attention has focused on the role of individual reovirus genes, it is clear that host factors contribute to reovirus tropism and pathogenesis. For example, preexisting hepatic injury alters the localization and extent of reovirus-induced hepatic disease (PICCOLI et al. 1990; RUBIN et al. 1990). The immunocompetence of the mouse also affects the expression of reovirus-induced disease. Adult immunocompetent mice are resistant to lethal infection, but adult SCID mice are killed by reovirus (TARDIEU et al. 1983; GEORGE et al. 1990; HALLER et al. 1995a,b). Specific types of immune responses can alter viral tropism and disease phenotype. For example, protection against lethal T3D-induced encephalitis by adoptive transfer of σ1-specific monoclonal antibody 9BG5 is associated with prevention of cortical destruction, but late development of striking hippocampal inflammation (TYLER et al. 1989).

Intramuscular or intracranial inoculation of neonatal mice with T3D regularly generates viral titers of 10^8–10^9 PFU per brain. In contrast, titers in the brain of adult SCID mice, despite their lack of functional immunity, range from 10^3–10^4 PFU per brain (GEORGE et al. 1990; HALLER et al. 1995b). While neonatal mice infected with serotype 3 reoviruses die of severe meningoencephalitis, adult SCID mice develop hepatitis or myocarditis (GEORGE et al. 1990; SHERRY et al. 1993; HALLER et al. 1995b). These data strongly argue that factors such as innate immunity and tissue maturation are important determinants of reovirus tropism.

2.2 Potential Role of the Immune System in Determining Organ-Specific Virulence Genes

One of the important findings from analysis of the roles of reovirus gene segments in multiple models of disease is the fact that different genome segments are important in different organs, a finding which supports the concept of organ-specific virulence genes (Fig. 1). For example, in the adult SCID mouse model, genome segments L1, L2, and S1 are important determinants of virulence and growth in

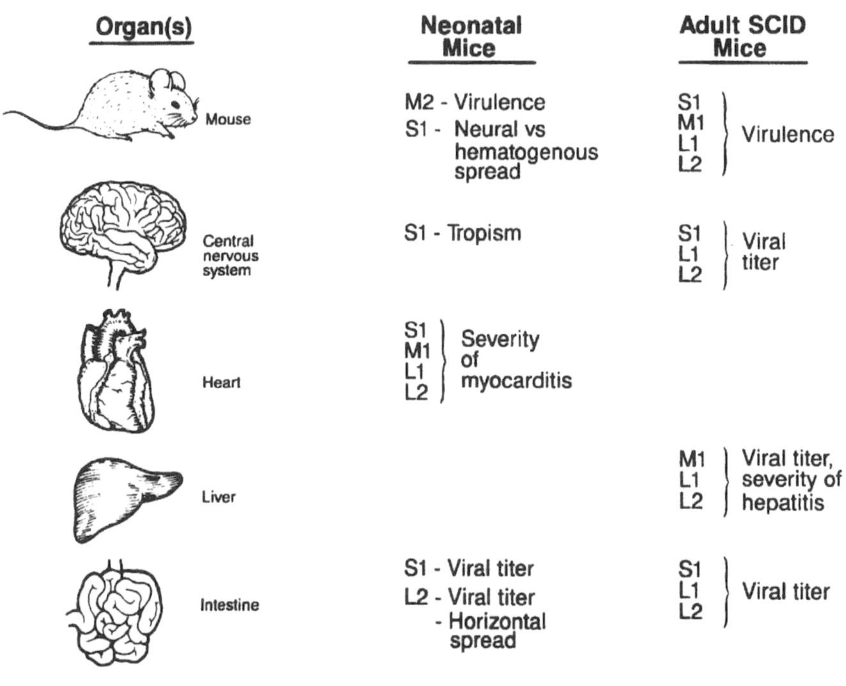

Fig. 1. Organ-specific virulence genes. Genome segments demonstrated to be important for either growth or tropism in the organ schematically depicted are shown

brain and intestine. No role for the M1 segment is detected in brain or intestine despite the use of sensitive statistical techniques (HALLER et al. 1995b). In contrast, viral yield as well as the number of hepatic inflammatory lesions in the liver of SCID mice is determined by segments L1, L2, and M1, while no role for the S1 segment is observed. Similarly, in neonatal mice the M1, L1, L2, and S1 segments play an important role in induction of myocarditis (SHERRY and FIELDS 1989; SHERRY and BLOOM 1994), while M2 and S1 are important in the CNS (see the chapter by K. TYLER, this volume). Similar to studies using adult SCID mice, L2 and S1 are important for controlling viral growth in the neonatal mouse intestine (BODKIN and FIELDS 1989; KEROACK and FIELDS 1986).

One model to explain the finding that different genome segments are important in specific organs is sequential involvement of different proteins in reovirus replication. In this model viral proteins vary in their importance in specific tissues because the rate limiting step in viral growth or clearance differs from organ to organ. For example, if the S1 segment encoded σ1 attachment protein is important for growth in CNS or intestine, then steps in viral replication related to attachment or entry might be rate-limiting in CNS or intestine. In the liver the M1 segment encoded μ2 protein would provide a rate-limiting function in viral replication or clearance, while S1 segment encoded proteins would not be rate limiting. Alternatively, the host response controlling viral replication might be distinct in different organs. Recent data from studies of herpesviruses shows that distinct parts of the immune response are important in different organs (TAY and WELSH 1997; JONJIC et al. 1989). This is consistent with the immune system playing a critical role in determining which viral genes are important in different organs. In this model, viral genome segments important in determining viral growth or tissue injury would be involved in steps in viral replication that are targeted by tissue-specific host factors (for example, cytokines).

3 Overview of Immunity to Reoviruses

While the pathogenesis of reovirus in multiple organs has been studied for many years, interest in reovirus immunity has recently intensified. This area has attracted attention because of the potential for integrating an understanding of functions of specific proteins at different stages in pathogenesis with information about protein-specific immunity. While a number of studies have evaluated activities of anti-reovirus antibodies, less is known about T cell recognition of reovirus antigens and the potential influence of T cells on reovirus pathogenesis. The role of interferons in reovirus resistance is dealt with in other chapters of this volume (see the chapter by C.E. SAMUEL, this volume). Innate responses including activation of natural killer cells, infiltration of neutrophils, and expression of cytokines such as tumor necrosis factor have been observed during reovirus infection, but their specific functional roles have not been defined (TATERKA et al. 1995; FARONE et al. 1996).

The reovirus system illustrates several important concepts about mechanisms of antibody action in vivo and in vitro. For example, studies using monoclonal antibodies clearly demonstrate that antibody can block neural spread of virus without affecting replication at the primary site of infection. Viral susceptibility to the action of antibody can be determined by viral proteins other than the antibody's target antigen. The lack of clear correlation between in vivo protective capacity of antibodies and their in vitro properties led to the discovery that some protective antibodies act by inhibiting steps in viral replication subsequent to cell attachment. Lastly, recent studies show that circulating IgG (independent of IgA) can play an important role in clearance of virus infection from the intestine.

Many studies have defined the in vitro activities of antibodies specific for various reovirus proteins, but until recently these studies contributed more information about the structure and function of reovirus proteins than the nature of protective immunity. In addition, because so much of reovirus biology is related to reovirus serotype, study of serotype-specific immune responses has overshadowed investigation of immune responses that are not serotype specific. As serotype specificity is primarily determined by epitopes on the σ1 protein, this focus has tended to emphasize immunity to σ1. More recently the presence of cross-reactivity between serotypes has attracted more attention, with consequent demonstration of protective responses to non-σ1 epitopes.

All currently recognized anti-reovirus immune responses are specific for protein antigens. There is no evidence for reovirus-specific anti-carbohydrate, anti-lipid, or anti-dsRNA immune responses. It is important to recognize that there is tremendous conservation of amino acid sequences in proteins from prototypic and field isolate reovirus strains. For example, the σ3 protein contains 365 amino acids, and T1L and T3D are different at only 11 amino acid positions. Thus it is not surprising that a significant component of both T and B cell immunity is not serotype-specific. For example, polyclonal antibodies raised against prototypic strain T1L precipitates T3D proteins (Gaillard and Joklik 1980), and even the serotype-defining σ1 protein has some serotype-cross-reactive epitopes (Gaillard and Joklik 1980; Hayes et al. 1981; Virgin et al. 1991).

4 Serotype-Specific and Cross-Reactive Recognition of Reovirus Antigens by Antibody

There are both serotype-specific and serotype-independent antigens on reoviruses (Rosen 1960). The availability of reassortant genetic techniques allowed rapid identification of targets of certain serotype-specific responses. Reassortant analysis shows that the S1 segment encoded σ1 protein is the primary determinant of serotype-specific neutralization (Weiner and Fields 1977). This conclusion is in agreement with the fact that the σ1 protein is more polymorphic than any other structural reovirus protein. Several studies show that the majority of neutralizing monoclonal antibodies specific for σ1 exhibit serotype-specific binding or neutralization of prototype reovirus strains (Hayes et al. 1981; Burstin et al. 1982; Virgin et al. 1988, 1991). In addition, monoclonal antibodies specific for T1L and T3D σ1 proteins show serotype-specific binding to a panel of nonprototype serotype 1 and serotype 3 reoviruses (Virgin et al. 1991). These observations support the use of σ1-specific monoclonal antibodies, such as 9BG5 and 5C6, for determining the serotype of reovirus isolates (Goral et al. 1996).

In contrast to σ1-specific monoclonal antibodies, monoclonal antibodies specific for other outer capsid proteins typically cross-react with proteins from prototype reoviruses of different serotypes (Hayes et al. 1981; Virgin et al. 1991). While these antibodies are cross-reactive, certain antibodies specific for the σ3, μ1, and λ2 capsid proteins consistently bound better to a panel of serotype 3 viruses than a panel of serotype 1 viruses (Virgin et al. 1991). This is similar to results of earlier experiments demonstrating that some monoclonal antibodies specific for proteins other than σ1 selectively precipitated proteins of one serotype more efficiently than other serotypes (Lee et al. 1981). Even polyclonal antibody raised against prototype reoviruses shows some level of serotype specificity in immunoprecipitating σ3, λ2, and μ1C (Gaillard and Joklik 1980). The apparent higher avidity of certain monoclonal antibodies to σ3, λ2, and μ1C from serotype 3 than serotype 1 viruses may reflect subtle differences in the conformation of outer capsid proteins in serotype 1 and serotype 3 viruses. Alternatively, the σ1 proteins of serotype 1 and serotype 3 reoviruses might differ in their interactions with outer capsid proteins σ3, μ1C, and λ2, causing changes in the conformation of these proteins and consequent serotype-dependent epitope recognition by monoclonal antibodies.

5 T Cell Responses to Reoviruses

T cell responses are induced during reovirus infection. This has been best demonstrated by evaluating changes in T cell populations in intestinal tissue after peroral or respiratory inoculation. After peroral challenge with T1L both Peyer's patches and intraepithelial lymphocyte preparations have CTL precursors that generate reovirus-specific CTL after in vitro stimulation (London et al. 1987, 1989a). These cells are MHC-restricted (London et al. 1987, 1990). Reovirus infection is also followed by development of IgA memory cells and both CD4 T cells and virus-specific helper cell responses (London et al. 1987, 1989a). After respiratory inoculation CD4 T cells, CD8 T cells, and plasma cells are observed in the inflammatory response (Thompson et al. 1996; Bellum et al. 1996). While CTL responses, and likely CD4 responses, to reovirus clearly occur, the physiological importance of these responses is less clear.

A number of studies show that immune cells protect against reovirus infection (Sherry et al. 1993; Virgin and Tyler 1991; Cuff et al. 1991). However, in mice with an interrupted β_2-microglobulin gene, which have a significant deficiency in

CD8 T cells, reovirus is cleared from the intestine with normal kinetics (BARKON et al. 1996). This finding suggests that normal levels of CTLs are not required for clearance of reovirus from the intestine, but does not rule out that these cells, or perhaps CD4 T cells, have some role in the normal response to reovirus (VIRGIN and TYLER 1991).

Few studies have been conducted to identify the protein antigens recognized by reovirus-specific T cells. In particular, newer methods to map peptides responsible for T cell recognition by MHC class I or class II restricted T cells have not been performed. However, reassortant genetic analysis has defined certain proteins as targets. Initial studies of CTL killing of reovirus-infected cells suggested that a prominent part of the CTL response to reovirus is serotype-specific (FINBERG et al. 1979, 1982) and directed to products of the S1 segment (σ1 or σ1s). While serotype-specific killing was observed, significant cross-reaction between targets infected with viruses of different serotypes was seen. Similar data demonstrating that serotype-specific T cell responses can be directed to S1 segment products were obtained during evaluation of the delayed type hypersensitivity response of BALB/c mice to rechallenge with reovirus (WEINER et al. 1980). Further analysis clearly showed that while serotype-specific CTL exist many reovirus-specific CTLs cross-react with viruses of different serotypes (HOGAN and CASHDOLLAR 1991; LONDON et al. 1989a,b). Thus there must be CTL epitopes on multiple reovirus proteins in addition to either σ1 or σ1s. Recently T cell specificity for the σ1NS nonstructural protein of reovirus has been observed, and that these responses include both strain-specific and cross-reactive components (HOFFMAN et al. 1996).

An anti-idiotypic monoclonal antibody raised against the T3D σ1-specific neutralizing monoclonal antibody 9BG5 can be used to immunize mice (SHARPE et al. 1984). The subsequent delayed-type hypersensitivity response is directed against σ1 (SHARPE et al. 1984). This same anti-idiotypic antibody elicited serotype-specific anti-reovirus antibody in several strains of mice, and mothers immunized with the anti-idiotype monoclonal antibody give birth to pups resistant to viral infection (GAULTON et al. 1986). The molecular basis of this cross-reactive recognition of viral epitopes by T cells and antibody specific for the anti-idiotypic antibody is likely related to the presence of sequences in the variable domain of the anti-idiotype with homology to sequences in the σ1 protein (WILLIAMS et al. 1988, 1989).

6 Neutralization of Reovirus Infectivity

Neutralization of reovirus by antibody is usually measured as plaque reduction (VIRGIN et al. 1991; HAYES et al. 1981), but the mechanism of neutralization has not been fully delineated for reovirus. Certain monoclonal antibodies (e.g., monoclonal antibody 9BG5, which is specific for T3D σ1) are efficient at plaque reduction neutralization and also efficiently block binding of virus to L cells (BURSTIN et al. 1982; VIRGIN et al. 1988, 1994). A number of very efficient neutralizing anti-σ1 monoclonal antibodies also aggregate virions (HAYES et al. 1981), but the role of virion aggregation in plaque reduction neutralization has not been completely defined. While serotype-specific neutralization can be observed using polyclonal sera, polyclonal anti-T1L serum efficiently neutralizes T3D and vice-versa (HAYES et al. 1981; VIRGIN et al. 1988). Monoclonal antibodies specific for σ1 often have plaque reduction activity (VIRGIN et al. 1991; HAYES et al. 1981; BURSTIN et al. 1982), as do some monoclonal and monospecific antibodies to λ2 and σ3 (HAYES et al. 1981). Mechanisms responsible for plaque-reduction neutralization by anti-σ3 and anti-λ2 monoclonal antibodies have not been defined.

7 Hemagglutination Inhibition and the Genetics of Viral Susceptibility to Antibody Action

Polyclonal anti-reovirus antibodies can inhibit virion-mediated hemagglutination (ROSEN 1960). This is likely the result, at least for antibodies specific for the viral hemagglutinin σ1, of antibody blocking the interaction between the virion and the erythrocyte surface. Monoclonal antibodies specific for σ3 and μ1 efficiently inhibit hemagglutination (VIRGIN et al. 1991; HAYES et al. 1981), which shows that antibody specific for one outer capsid protein (σ3) can inhibit the function of another outer capsid protein (σ1). The hemagglutination-inhibiting capacity of monoclonal anti-σ3 antibodies is dependent on the target virus. For example, monoclonal antibody 8F12 (specific for both T3D and T1L σ3) inhibits hemagglutination by T1L more efficiently than T3D, while monoclonal antibody 4F2 (specific for both T1L and T3D σ3) inhibits hemagglutination by T3D more efficiently than T1L. These differences are not explained by the avidity of anti-σ3 monoclonal antibodies for T1L versus T3D (VIRGIN et al. 1991), but instead may be due to strain-dependent interactions between σ1 and σ3. Reassortant genetic analysis showed that strain-specific differences in hemagglutination inhibition capacity of anti-σ3 monoclonal antibodies is determined by the S1 segment. Thus monoclonal antibody 4F2, which is specific for σ3, inhibits hemagglutination by reassortant viruses containing the T3D but not the T1L S1 segment (VIRGIN et al. 1991). This shows that the primary determinant of susceptibility to antibody action can be a protein other than the antibody's target antigen. This is an important concept since the efficacy of subunit vaccines may depend on viral proteins other than the vaccine antigen.

8 Adoptive Transfer of Protection Against Reovirus-Induced Disease

Antibody (both monoclonal and polyclonal, both serotype-specific and serotype cross-reactive) can protect against lethal infection with serotype 3 reoviruses (T3D and T3C9) (Tyler et al. 1989, 1993; Virgin et al. 1988; Cuff et al. 1990). In addition, antibody can protect against T1L-induced hydrocephalus (Tyler et al. 1993) and 8B-induced myocarditis (Sherry et al. 1993). Antibody has been used prophylactically to prevent lethal disease and as treatment for established CNS infection (Virgin et al. 1988). Pups born of immune dams are resistant to lethal challenge with reovirus (Cuff et al. 1990), and this protection appears to be related to both transplacental transfer of antibody and antibody in breast milk (Cuff et al. 1990). Immunization of dams with monoclonal anti-idiotype also protects pups from reovirus infection (Gaulton et al. 1986).

Adoptive transfer of immune spleen cells protects neonatal mice from lethal infection with either perorally administered T3C9 or intramuscularly administered T3D (Virgin and Tyler 1991). Protection against hydrocephalus caused by T1L and myocarditis caused by 8B also has been observed (Sherry et al. 1993; Virgin and Tyler 1991). In this adoptive transfer model both CD4 and CD8 T cells are required for maximal protection against T3D, T3C9, or 8B (Sherry et al. 1993; Virgin and Tyler 1991). Further evidence for a role of both CD4 and CD8 T cells in protection against reovirus-induced disease comes from experiments using monoclonal antibodies to deplete CD4 and CD8 T cells in vivo (Virgin and Tyler 1991). In these experiments, depletion of CD4 and CD8 T cells increased the severity and changed the nature of T3C9-induced liver disease. Adoptive transfer of Peyer's patch cells protects adult SCID mice against lethal infection (George et al. 1990) and neonatal mice against peroral challenge with T3C9 (Cuff et al. 1991). This shows that protective immune cells are present in the intestine, which is the primary portal of viral entry into the host.

9 Role of Antibody at Defined Pathogenetic Stages: Spread Within the Host

The effect of antibody at different stages in reovirus pathogenesis has been evaluated in detail (Fig. 2), based on extensive studies defining the stages in reovirus pathogenesis. Monoclonal antibodies specific for outer capsid proteins σ1 (Tyler et al. 1989, 1993; Virgin et al. 1988), σ3 (Tyler et al. 1993), and μ1 (Tyler et al. 1993) can be protective in vivo. One prominent in vivo mechanism of antibody action is inhibition of neural spread of virus from primary sites in muscle and intestine to the nervous system. T3D spreads from hind-limb muscle to the CNS via the sciatic nerve while T3C9 spreads from the intestine to the CNS via the vagus nerve. Monoclonal antibody-mediated protection against T3D-induced disease after intramuscular inoculation is associated with inhibition of spread to the inferior spinal cord, in some cases without effects on primary replication (Tyler et al. 1989, 1993).

In addition, viral spread within the CNS, for example, from the brain to the retina or from the inferior to the superior spinal cord, is inhibited by adoptively

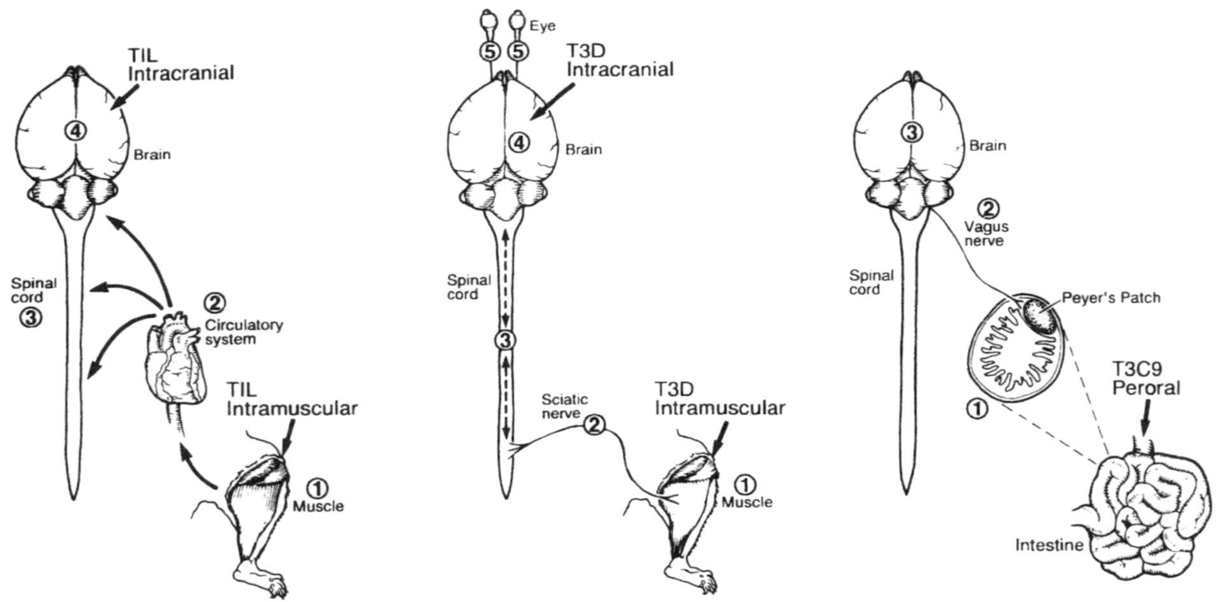

Fig. 2. Role of antibody at specific stages in pathogenesis of reovirus infection. *Numbers*, sites at which antibody has specific functions. For hematogenous spread after intramuscular or intracranial inoculation of T1L (*left*), antibody can: (*1*) inhibit replication in muscle, (*2*) decrease viremia independent of effects on replication in muscle, (*3*) inhibit spread to and/or growth in the spinal cord, (*4*) inhibit spread to and/or growth in the brain, and prevent development of hydrocephalus. For neural spread after intramuscular or intracranial inoculation of T3D (*middle*) antibody can: (*1*) inhibit replication in muscle, (*2*) inhibit neural spread to and/or growth in the spinal cord independent of effects on replication in muscle, (*3*) inhibit spread to and/or growth in and spread via the spinal cord, (*4*) inhibit growth in the brain and prevent lethal encephalitis, and (*5*) inhibit neural spread to and/or growth in the eye from the brain. For neural spread after peroral inoculation of T3C9 (*right*), antibody can: (*1*) enhance intestinal clearance of virus after peroral inoculation of adult B cell deficient mice. However, circulating IgG does not effect primary replication early (days 3–5) after peroral inoculation of neonatal mice, (*2*) inhibit neural spread to and/or growth in the brain via the vagus nerve, and (*3*) inhibit growth in the brain and prevent lethal encephalitis

transferred antibody (TYLER et al. 1989). Consistent with these findings, monoclonal antibodies inhibit spread of T3C9 from the intestine via the vagus nerve to the brain, in many cases without altering T3C9 titer in intestine (TYLER et al. 1989, 1993). Thus in two distinct disease models antibody inhibits neural spread from primary sites of infection when effects on primary replication are minimal or absent. Mechanisms by which antibody inhibits neural spread independent of replication at the primary site of inoculation have not been defined.

10 Role of Antibody at Defined Pathogenetic Stages: Effects on Primary Replication in the Intestine

While the capacity of antibody to protect against systemic disease with reovirus and the capacity of adoptively transferred immune cells to control intestinal virus replication are clear, less is known about mechanisms responsible for controlling viral replication and finally clearing virus from primary sites of infection. Certain monoclonal antibodies inhibit viral replication after intramuscular inoculation with T3D, establishing that antibodies can act at primary sites (TYLER et al. 1993). The importance of B cells and antibody at a natural portal of entry, the intestine, has been addressed in a recent study comparing the clearance of reovirus T3C9 from the intestine after peroral inoculation of adult immunocompetent and immunocompromised mice. SCID mice, and mice with targeted interruptions of either the transmembrane exon of IgM (B cell and antibody deficient, KITAMURA et al. 1991) or the β_2-microglobulin component of MHC class I molecules (CD8 T cell deficient, KOLLER et al. 1990) were compared (BARKON et al. 1996). SCID mice failed to clear intestinal infection after peroral inoculation, while immunocompetent mice cleared virus within 7 days.

This confirms that functional lymphocytes are important for clearance of virus from the intestine (GEORGE et al. 1990; HALLER et al. 1995b). Despite the prominent CD8 CTL response documented in intestine after peroral inoculation of reovirus, mice with an interrupted β_2-microglobulin gene cleared T3C9 from the intestine with normal kinetics. In contrast, mice deficient in B cells and antibody failed to control T3C9 replication in intestine between 7 and 11 days after inoculation, demonstrating an important role for B cells in intestinal clearance. This was confirmed in studies showing that adoptive transfer of immune B cells into SCID mice allowed control of intestinal infection. Additionally, systemic administration of IgG specific for reovirus reconstituted the ability of B cell deficient mice to clear primary infection from the intestine (BARKON et al. 1996), showing that circulating IgG can play a protective role at mucosal surfaces independent of IgA.

11 Effect of Antibody on Virus-Cell Interactions

While a number of activities of antibody in vitro have been examined, recent studies have focused on the role of antibodies after virus-antibody complexes bind to the cell. These studies were undertaken because, in the reovirus system, antibody-mediated protection is not consistently correlated with in vitro properties of antibodies including antibody isotype, antibody avidity, plaque-reduction neutralization, hemagglutination inhibition, or inhibition of viral binding to target cells (TYLER et al. 1989, 1993; VIRGIN et al. 1988). For example, several monoclonal antibodies specific for the σ3 outer capsid protein are nonneutralizing but are protective in vivo (TYLER et al. 1993). Monoclonal antibody-mediated protection against T3D-induced CNS disease is seen in mice depleted of serum complement (VIRGIN et al. 1988), arguing against a critical role for complement in protection by antibody. Thus the available data in the reovirus system strongly argue against reliance on in vitro properties of antibody (such as neutralization) as predictors of in vivo efficacy of antibody responses.

These findings led to a search for other actions of antibodies that might explain their protective capacity (Fig. 3). It was observed that antibodies specific for σ3 and μ1 inhibited replication of virus already bound to the cell surface (VIRGIN et al. 1994). This effect on postbinding events is a consistent predictor of in vivo pro-

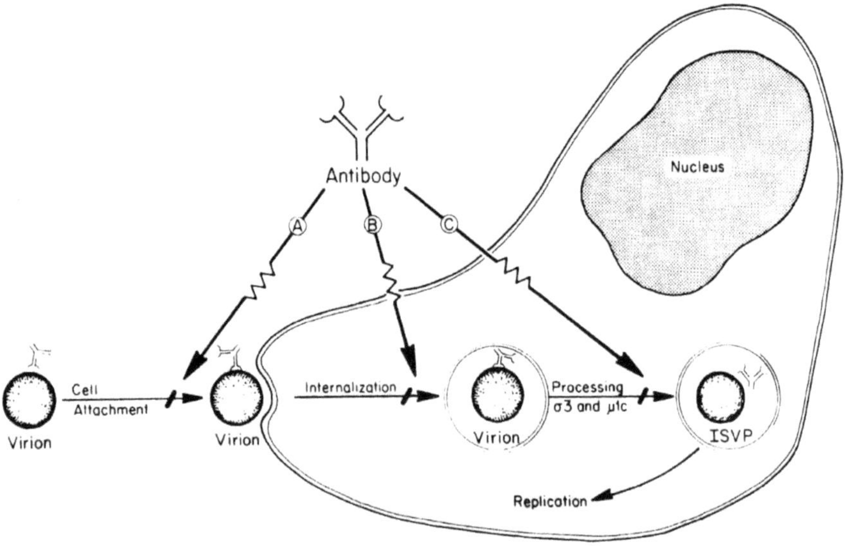

Fig. 3. Effects of antibody on virus-cell interactions. Antibodies specific for virion outer capsid proteins inhibit multiple stages in the interaction between reoviruses and the cell. *A*, Some antibodies specific for the 1 protein block cell attachment of virus to the cell; *B*, an antibody specific for the 3 protein, when bound to the virus prior to virus binding to the cell, inhibits internalization of the virus; *C*, antibodies specific for the 3 protein, when bound to the virus before the virus binds to the cell, block the intracellular proteolytic uncoating of the virion to the ISVP form. (With permission from VIRGIN et al. 1994)

tective capacity (Tyler et al. 1993; Virgin et al. 1994). Multiple mechanisms contribute to antibody-mediated inhibition of viral growth (Fig. 3). First, some monoclonal antibodies inhibit reovirus internalization. In addition, monoclonal anti-σ3 antibodies inhibit intracellular proteolytic uncoating of the reovirus outer capsid, and augments the action of ammonium chloride (which also inhibits virion uncoating) (Virgin et al. 1994). This is an effect of antibody on proteolysis of σ3 since antibodies also blocked uncoating of virions by chymotrypsin in vitro. Antibodies specific for the μ1 outer capsid protein inhibit membrane penetration by ISVPs, suggesting an additional intracellular site of antibody action (Hooper and Fields 1996). These studies suggest that subunit vaccine design might be effectively directed to epitopes critical for postbinding steps in viral replication, and that antiviral drugs might be combined with antibodies targeted to the same steps in viral replication.

12 Conclusion

The reovirus model system has a place in modern immunological studies as a system in which structural information, cell biological studies, and genetic approaches can be used to direct and complement immunological studies. These advantages validate this model for immunological studies despite the fact that there is no direct human disease connection. The finding of organ-specific virulence genes, the identification of viral genes that regulate susceptibility of the virus to antibody action, the definition of intracellular actions of antibody, and the detailed information derived from analysis of how antibody functions in the intestine and nervous system are clear examples of the advantages of immunological studies in this system. Studies that take advantage of the reovirus system (most especially genetic approaches) to probe innate responses, T cell responses, and the role of cytokines will likely yield new and interesting information.

References

Barkon ML, Haller BL, Virgin HW (1996) Circulating IgG can play a critical role in clearance of intestinal reovirus infection. J Virol 70:1109–1116

Bellum SC, Hamamdzic D, Thompson AH, Harley RA, London SD, London L (1996) Experimental reovirus serotype 1/strain Lang infection of the lung: a model for the study of the lung in the context of mucosal immunity. Lab Invest 74:221–231

Bodkin DK, Fields BN (1989) Growth and survival of reovirus in intestinal tissue: role of the L2 and S1 genes. J Virol 63:1188–1193

Burstin SJ, Spriggs DR, Fields BN (1982) Evidence for functional domains on the reovirus type 3 hemagglutinin. Virology 117:146–155

Cuff CF, Lavi E, Cebra CK, Cebra JJ, Rubin DH (1990) Passive immunity to fatal reovirus serotype 3-induced meningoencephalitis mediated by both secretory and transplacental factors in neonatal mice. J Virol 64:1256–1263

Cuff CF, Cebra CK, Lavi E, Molowitz EH, Rubin DH, Cebra JJ (1991) Protection of neonatal mice from fatal reovirus infection by immune serum and gut derived lymphocytes. Adv Exp Med Biol 310:307–315
Farone AL, Frevert CW, Farone MB, Morin MJ, Fields BN, Paulauskis JD, Kobzik L (1996) Serotype-dependent induction of pulmonary neutrophilia and inflammatory cytokine gene expression by reovirus. J Virol 70:7079–7084
Finberg R, Weiner HL, Fields BN, Benacerraf B, Burakoff SJ (1979) Generation of cytolytic T lymphocytes after reovirus infection: role of S1 gene. Proc Natl Acad Sci USA 76:442–446
Finberg R, Spriggs DR, Fields BN (1982) Host immune response to reovirus: CTL recognize the major neutralization domain of the viral hemagglutinin. J Immunol 129:2235–2238
Gaillard RK, Joklik WK (1980) The antigenic determinants of most of the proteins coded by the three serotypes of reovirus are highly conserved during evolution. Virology 107:533–536
Gaulton GN, Sharpe AH, Chang DW, Fields BN, Greene MI (1986) Syngeneic monoclonal internal image anti-idiotopes as prophylactic vaccines. J Immunol 137:2930–2936
George A, Kost SI, Witzleben CL, Cebra JJ, Rubin DH (1990) Reovirus-induced liver disease in severe combined immunodeficient (SCID) mice. A model for the study of viral infection, pathogenesis, and clearance. J Exp Med 171:929–934
Goral MI, Mochow-Grundy M, Dermody TS (1996) Sequence diversity within the reovirus S3 gene: reoviruses evolve independently of host species, geographic locale, and date of isolation. Virology 216:265–271
Haller BL, Barkon ML, Li X, Hu WM, Wetzel JD, Dermody TS, Virgin HW (1995a) Brain and intestine-specific variants of reovirus serotype 3 strain Dearing are selected during chronic infection of severe combined immunodeficient mice. J Virol 69:3933–3937
Haller BL, Barkon ML, Vogler G, Virgin HW (1995b) Genetic mapping of reovirus virulence and organ tropism in severe combined immunodeficient mice: organ specific virulence genes. J Virol 69:357–364
Hayes EC, Lee PW, Miller SE, Joklik WK (1981) The interaction of a series of hybridoma IgGs with reovirus particles: demonstration that the core protein lambda 2 is exposed on the particle surface. Virology 108:147–155
Hoffman LM, Hogan KT, Cashdollar LW (1996) The reovirus nonstructural protein 1NS is recognized by murine cytotoxic lymphocytes. J Virol 70:8160–8164
Hogan KT, Cashdollar LW (1991) Clonal analysis of the cytotoxic T-lymphocyte response to reovirus. Viral Immunol 4:167–175
Hooper JW, Fields BN (1996) Monoclonal antibodies to reovirus sigma 1 and mu 1 proteins inhibit chromium release from mouse L cells. J Virol 70 (1):672–677
Jonjic S, Mutter W, Weiland F, Reddehase MJ, Koszinowski UH (1989) Site-restricted persistent cytomegalovirus infection after selective long-term depletion of CD4+ T lymphocytes. J Exp Med 169:1199–1212
Keroack M, Fields BN (1986) Viral shedding and transmission between hosts determined by reovirus L2 gene. Science 232:1635–1638
Kitamura D, Roes J, Kuhn R, Rajewsky K (1991) A B cell-deficient mouse by targeted disruption of the membrane exon of the immunoglobulin mu chain gene. Nature 350:423–426
Koller BH, Marrack P, Kappler JW, Smithies O (1990) Normal development of mice deficient in beta 2 M, MHC class I proteins, and CD8+ T cells. Science 248:1227–1230
Lee PW, Hayes EC, Joklik WK (1981) Characterization of anti-reovirus immunoglobulins secreted by cloned hybridoma cell lines. Virology 108:134–146
London SD, Rubin DH, Cebra JJ (1987) Gut mucosal immunization with reovirus serotype 1/L stimulates virus-specific cytotoxic T cell precursors as well as IgA memory cells in Peyer's patches. J Exp Med 165:830–847
London SD, Cebra JJ, Rubin DH (1989a) Intraepithelial lymphocytes contain virus-specific, MHC-restricted cytotoxic cell precursors after gut mucosal immunization with reovirus serotype 1/Lang. Reg Immunol 2:98–102
London SD, Cebra JJ, Rubin DH (1989b) The reovirus-specific cytotoxic T cell response is not restricted to serotypically unique epitopes associated with the virus hemagglutinin. Microb Pathog 6:43–50
London SD, Cebra-Thomas JA, Rubin DH, Cebra JJ (1990) CD8 lymphocyte subpopulations in Peyer's patches induced by reovirus serotype 1 infection. J Immunol 144:3187–3194
Morrison LA, Fields BN (1991) Parallel mechanisms in neuropathogenesis of enteric virus infections. J Virol 65:2767–2772
Nibert ML, Furlong DB, Fields BN (1991) Mechanisms of viral pathogenesis. Distinct forms of reoviruses and their roles during replication in cells and host. J Clin Invest 88:727–734

Nibert ML, Schiff LA, Fields BN (1995) Reoviruses and their replication. In: Fields BN, Knipe DM (eds) Virology. Raven, New York
Piccoli DA, Witzleben CL, Guico CJ, Morrison A, Rubin DH (1990) Synergism between hepatic injuries and a nonhepatotropic reovirus in mice. Enhanced hepatic infection and death. J Clin Invest 86:1038–1045
Rosen L (1960) Serologic groupings of reovirus by hemagglutination inhibition. Am J Hyg 71:242–249
Rubin DH, Morrison AH, Witzleben CL, Guico CJ, Piccoli DA (1990) Site of reovirus replication in liver is determined by the type of hepatocellular insult. J Virol 64:4593–4597
Sharpe AH, Gaulton GN, McDade KK, Fields BN, Greene MI (1984) Syngeneic monoclonal antiidiotype can induce cellular immunity to reovirus. J Exp Med 160:1195–1205
Sherry B, Bloom MA (1994) Multiple viral core proteins are determinants of reovirus-induced acute myocarditis. J Virol 68:8461–8465
Sherry B, Fields BN (1989) The reovirus M1 gene, encoding a viral core protein, is associated with the myocarditic phenotype of a reovirus variant. J Virol 63:4850–4856
Sherry B, Li X-Y, Tyler KL, Cullen JM, Virgin HW (1993) Lymphocytes protect against and are not required for reovirus induced myocarditis. J Virol 67:6119–6124
Tardieu M, Powers ML, Weiner HL (1983) Age dependent susceptibility to reovirus type 3 encephalitis: role of host and viral factors. Ann Neurol 13:602–607
Taterka J, Cebra JJ, Rubin DH (1995) Characterization of cytotoxic cells from reovirus-infected SCID mice: activated cells express natural killer- and lymphokine activated killer-like activity but fail to clear infection. J Virol 69:3910–3914
Tay CH, Welsh RM (1997) Distinct organ-dependent mechanisms for the control of murine cytomegalovirus infection by natural killer cells. J Virol 71:267–275
Thompson AH, London L, Bellum SC, Hamamdzic D, Harley RA, London SD (1996) Respiratory-mucosal lymphocyte populations induced by reovirus serotype 1 infection. Cell Immunol 169:278–287
Tyler KL, Fields BN (1995) Reoviruses. In: Fields BN, Knipe DM (eds) Virology. Raven, New York
Tyler KL, Virgin HW, Bassel Duby R, Fields BN (1989) Antibody inhibits defined stages in the pathogenesis of reovirus serotype 3 infection of the central nervous system. J Exp Med 170:887–900
Tyler KL, Mann MA, Fields BN, Virgin HW (1993) Protective anti-reovirus monoclonal antibodies and their effects on viral pathogenesis. J Virol 67:3446–3453
Virgin HW, Tyler KL (1991) Role of immune cells in protection against and control of reovirus infection in neonatal mice. J Virol 65:5157–5164
Virgin HW, Bassel Duby R, Fields BN, Tyler KL (1988) Antibody protects against lethal infection with the neurally spreading reovirus type 3 (Dearing). J Virol 62:4594–4604
Virgin HW, Mann MA, Fields BN, Tyler KL (1991) Monoclonal antibodies to reovirus reveal structure/function relationships between capsid proteins and genetics of susceptibility to antibody action. J Virol 65:6772–6781
Virgin HW, Mann MA, Tyler KL (1994) Protective antibodies inhibit reovirus internalization and uncoating by intracellular proteases. J Virol 68:6719–6729
Virgin HW, Tyler KL, Dermody TS (1997) Reovirus. In: Nathanson N (ed) Viral pathogenesis. Lippincott-Raven, Philadelphia, pp 669–699
Weiner HL, Fields BN (1977) Neutralization of reovirus: the gene responsible for the neutralization antigen. J Exp Med 146:1305–1310
Weiner HL, Greene MI, Fields BN (1980) Delayed hypersensitivity in mice infected with reovirus I Identification of host and viral gene products responsible for the immune response. J Immunol 125:278–282
Williams WV, Guy HR, Rubin DH, Robey F, Myers JN, Kieber Emmons T, Weiner DB, Greene MI (1988) Sequences of the cell-attachment sites of reovirus type 3 and its anti-idiotypic/antireceptor antibody: modeling of their three-dimensional structures. Proc Natl Acad Sci USA 85:6488–6492
Williams WV, London SD, Weiner DB, Wadsworth S, Berzofsky JA, Robey F, Rubin DH, Greene MI (1989) Immune response to a molecularly defined internal image idiotope. J Immunol 142:4392–4400

Mucosal Immunity to Reovirus Infection

A.S. MAJOR[1], D.H. RUBIN[2] and C.F. CUFF[3]

1 Introduction

The immune systems of the gastrointestinal, respiratory, and urogenital tracts are the first lines of defense against viral pathogens that infect via the host's mucosal surfaces. Antigen-specific anti-viral mucosal immunity includes both humoral and cellular components which consist of virus-specific IgA (and to a lesser extent IgM and IgG), T helper (Th), and cytotoxic T lymphocytes (CTL). An understanding of viral pathogenesis and immunity at mucosal surfaces is important when considering approaches to developing protective and efficacious vaccines and delivery systems for gene therapy. However, the tedious nature of isolating immune cells from mucosal tissues and the lack of animal models for many viral diseases presents unique challenges to viral immunologists.

[1]Department of Microbiology and Immunology, Robert C. Byrd Health Sciences Center, West Virginia University School of Medicine, Morgantown, WV 26506, USA
[2]Research Medicine, Veterans Affairs Medical Center, and Departments of Medicine, Microbiology and Immunology, Vanderbilt University, Nashville, TN 37212, USA
[3]Department of Microbiology and Immunology, Robert C. Byrd Health Sciences Center, West Virginia University School of Medicine, Morgantown, WV 26506, USA

Reoviruses were initially isolated from mammalian gastrointestinal and respiratory tracts, but their association with clinical disease in man is uncertain. Reovirus infections in adult immunocompetent mice also do not usually lead to pathology unless they are given in doses greater that 10^9 PFU (Rubin et al. 1985, 1986). However, reovirus infection in neonatal and immunocompromised mice can result in gastrointestinal, hepatic, and central nervous system (CNS) disease (Stanley and Joske 1975; Margolis et al. 1971). In addition, reovirus serotype 1 strain Lang (T1L) and serotype 3 strain Dearing (T3D) can induce diabetes mellitus in mice (Onodera et al. 1981). Therefore reovirus has served as a model to study basic mechanisms of viral pathogenesis that are reviewed in other chapters. This chapter reviews the use of reovirus infection in immunocompetent and immunodeficient rodents as a means to probe the host's ability to induce, maintain, and regulate the mucosal immune responses of the gastrointestinal and respiratory tracts.

2 Reovirus Infection as a Model to Study Mucosal Immunity

Rubin et al. (1985, 1986) observed that mice infected with reovirus T1L by either the oral or intravenous routes developed intestinal pathology that is localized mostly to the ileum and consists of inflammation of the lamina propria, crypt-cell hyperplasia, and villi shortening. A model of reovirus infection has been proposed in which reovirus enters the host via the M cell prior to adsorption to receptors on the basolateral surface of epithelial cells (Wolf et al. 1981, 1987; Bass et al. 1988; see Fig. 1 in Organ and Rubin, this volume). Therefore reovirus is initially concentrated at the mucosal equivalent of a peripheral lymph node, Peyer's patches, ensuring high exposure of virus with mucosal immune cells. This specific association of reovirus with the mucosal immune compartment has made reovirus an effective probe of immune responses.

3 Humoral Immunity in Adult Mice

It is commonly thought that IgA is important in protection and prevention against viral infections at mucosal surfaces by binding to virus and either interfering with virus–host cell binding or virus adsorption onto cells (Outlaw and Dimmock 1990, 1991; Outlaw et al. 1990; Dimmock 1993). Experimental evidence also suggests that polyimmunoglobulin receptor-mediated uptake of virus-specific IgA into epithelial cells facilitates the intracellular inhibition of viral replication (Mazanec et al. 1992). One could surmise that production of IgA antibodies at mucosal sites would be advantageous to the host because IgA typically does not mediate inflammatory responses (Kilian and Russell 1994). Therefore development of inflammation

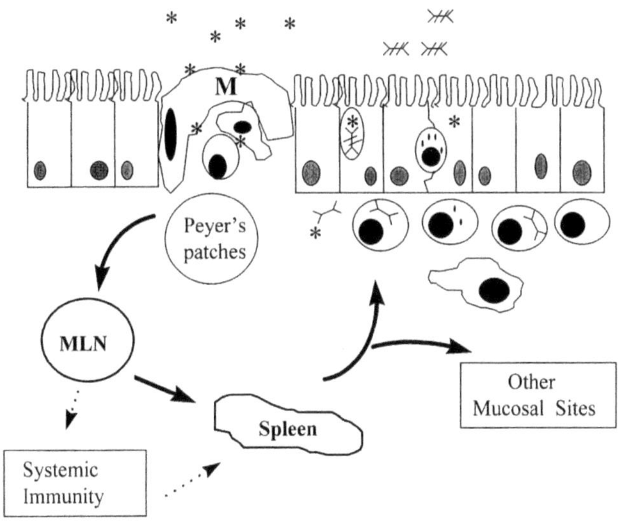

Fig. 1. Induction and expression of reovirus-specific intestinal immunity. Reovirus (T1L) binds to M cells and virus is translocated into the Peyer's patches. Virus-specific T cells and B cells are primed in the patch and migrate via efferent lymphatics to the mesenteric lymph node and spleen. From the spleen primed effectors migrate via the blood stream to other mucosal sites and back to the intestine, where they lodge in the lamina propria below the absorptive epithelium. Fully differentiated plasma cells secrete antibody which is translocated through epithelium to the intestinal lumen in the final form of secretory IgA. In addition, virus-specific pCTL develop in the IEL populations, presumably as a result of priming in the Peyer's patches. It is likely, although still unproved, that virus-specific T-helper and T-cytotoxic cells can be found in the lamina propria following infection. Immunological effectors mediated by IgG and T cells can also be found in the periphery following oral infection

and/or immunopathology that is associated with other antibody responses such as IgG and IgE is reduced in the intestines following antigenic challenge.

Intraduodenal injection of adult mice with reovirus T1L primes virus-specific IgA-producing cells in the Peyer's patches (RUBIN et al. 1983). Further studies revealed that the frequency of reovirus-specific IgA producing cells in Peyer's patches and spleens are higher after intraduodenal injection than following intraperitoneal injection (LONDON et al. 1987). These experiments support the hypothesis that virus-specific IgA responses are initiated in the Peyer's patches and suggest that preferential entry into the gut via M cells is important in eliciting mucosal humoral immunity. In addition, monoclonal IgA antibodies to σ3 and μ1c have been shown to bind to M cells (WELTZIN et al. 1989). Immune complexes of reovirus and IgA may therefore have enhanced uptake by M cells, resulting in efficient mucosal priming.

In addition to mucosal IgA, enteric reovirus infections in mice elicit virus-specific serum IgG. The majority of the reovirus-specific serum IgG is IgG2a and IgG2b, while levels of virus-specific IgG1 are low (WEINSTEIN and CEBRA 1991; MAJOR and CUFF 1996). Reovirus-specific IgG1 is produced following systemic infection but is dependent on the genetic background of the infected mouse (MAJOR and CUFF 1996).

Much of the reovirus-specific IgG is directed against the major structural proteins such as σ1, σ3, λ2, and μ1 (HAYES et al. 1981; VIRGIN et al. 1991; TYLER et al. 1993; WELTZIN et al. 1989). In vitro neutralization and inhibition studies using monoclonal antibodies against reovirus T3D indicated that virus-specific IgG may prevent viral infection at three definitive steps during reovirus replication (VIRGIN et al. 1994). Antibodies specific for the σ1 viral hemagglutinin protein are neutralizing and may inhibit virus/cell interaction, whereas antibodies directed toward epitopes on other viral proteins do not neutralize virus in tissue culture assays. Additionally, IgG specific for the σ3 and possibly μ1 outer capsid proteins are hypothesized to prevent penetration of virus into cells and prevent protease-mediated uncoating of virus. Thus "nonneutralizing" reovirus-specific IgG monoclonal antibodies can prevent in vivo infection. Although unknown, it is possible that IgA antibodies specific for determinants other than σ1 mediate similar events at mucosal surfaces.

In addition to being produced in systemic lymphoid tissue and found in serum, antigen-specific IgG has also been detected at mucosal sites. However, the role of intestinal IgG during reovirus infection remains somewhat unclear. Passive transfer of reovirus T3D-specific IgG to severe combined immunodeficient (SCID) mice and immunoglobulin-knockout mice facilitates reovirus clearance from intestines in the absence of virus-specific IgA (BARKON et al. 1996). However, in immunocompetent animals reovirus-specific IgG production in Peyer's patches and the lamina propria is virtually undetectable by ELISA after enteric reovirus infection (WEINSTEIN and CEBRA 1991; MAJOR and CUFF 1996). In addition, systemic administration of reovirus-specific IgG to immunocompetent neonatal mice fails to inhibit viral replication in the intestine (TYLER et al. 1989). This may reflect differences in virus uptake into epithelial cells in neonatal mice that is not present after gut closure (see ORGAN and RUBIN, this volume). Therefore in normal neonatal and adult mice reovirus-specific IgA seems to prevent intestinal spread at gut mucosal surfaces and virus-specific IgG appears to be important in preventing viral dissemination.

In studies of human exposure most young adults tested were seropositive for anti-reovirus antibodies (LERNER et al. 1947; JACKSON et al. 1961; LEERS and ROYCE 1966). SELB and WEBER (1994) demonstrated that virus-specific serum IgG but not IgA increases with age. Serum IgA peaked at various ages, which led to the suggestion that high levels of reovirus-specific serum IgA in humans are a marker for a recent infection. Immunoblot analyses of both human and mouse antibody demonstrate that the reovirus-specific IgA and IgG responses are polyclonal and specific for the λ, μ, and σ proteins (WELTZIN et al. 1989; SELB and WEBER 1994; MAJOR and CUFF 1996).

4 Cellular Immunity in Adult Mice

Reovirus-specific CTL were first described by FINBERG et al. (1979) following intraperitoneal infection. Reovirus T1L- and T3D-specific CTL could be expanded in vitro from murine splenocytes as early as 7 days and as late as 2 months after infection. Preferential in vivo depletion of major histocompatibility complex (MHC) class II bearing antigen presenting cells by UV irradiation led to decreases in reovirus-specific CTL responses in intraperitoneally infected mice (LETVIN et al. 1981). These data suggested that the generation of reovirus-specific CTL is T cell-dependent because Th cells that recognize MHC class II antigens were necessary to prime cytotoxic cells. Although these studies increased the understanding of reovirus immunity in systemically infected mice, it was not known whether similar cellular immunity existed in gut-associated lymphoid tissue (GALT).

LONDON et al. (1987) showed that virus-specific precursor CTL (pCTL) can be isolated from Peyer's patches 6 days after intraduodenal priming with active reovirus T1L. The cytotoxic activity of the reovirus-specific CTL was enhanced by the addition of IL-2 in culture media, indicating that Th cell activity is also involved in the generation of reovirus-specific CTL in the gut.

Treatment of reovirus-primed Peyer's patch lymphocytes with antibodies against Thy-1 or CD8 and complement inhibited specific lysis of infected cells, suggesting that the CTL were Thy-1^+ and $CD8^+$ T cells (LONDON et al. 1987). Reovirus-specific pCTL isolated from enterically primed Peyer's patches also expressed the germinal center T cell (GCT) surface marker (LONDON et al. 1990). Expression of GCT on Peyer's patch $CD8^+$ T cells was apparent by day 2 and peaked at day 6 after reovirus infection. Virus-specific cell lysis was inhibited when reovirus-primed Peyer's patch lymphocytes were treated with antibody against GCT and complement, demonstrating that GCT expression could be used as a marker for activation of reovirus-specific pCTL.

Following intraduodenal reovirus infection, Peyer's patch and mesenteric lymph node lymphocytes restimulated in vitro expressed greater virus-specific CTL activity than lymphocytes isolated from the spleens and peripheral lymph nodes (LONDON et al. 1987). Furthermore, limiting-dilution analyses revealed that the frequencies of reovirus-specific pCTL per 10^6 $CD8^+$ lymphocytes are 100-fold higher in Peyer's patches than in peripheral lymph nodes 6 days after enteric infection. Therefore the authors concluded that the reovirus-specific CTL response is initiated in Peyer's patches and is not due to virus priming of peripheral lymphoid tissue.

Intraduodenal infection of reovirus T1L also elicits specific pCTL among intraepithelial lymphocytes (IEL) (LONDON et al. 1989b). This was the first indication that virus-specific pCTL could be isolated from the intestinal epithelium (Table 1). Flow cytometric analyses and depletion experiments using antibodies specific for CD8 and Thy-1 demonstrated that reovirus-specific CTL that are expanded in culture were α/β T cell Receptor (TCR^+), $CD8^+$ T cells. Adoptive transfer of Peyer's patch lymphocytes from reovirus-nonimmune, immunocompetent mice to reovirus-infected SCID mice resulted in the appearance of pCTL in the intestinal epithelium (CUFF et al. 1993). This suggested that the Peyer's patch pCTL are developmentally related to virus-specific pCTL isolated in the IEL. In addition, analyses of IEL from germ-free mice enterically infected with reovirus T1L indicated that the number of IEL are markedly increased after infection, and that the predominant phenotype shifts from γ/δ TCR^+ to α/β TCR^+ T cells (CUFF et al.

Table 1. Virus-specific MHC-restricted CTL activity by IEL from reovirus-infected mice (adapted from LONDON et al. (1989b)

Effector	Target	% Specific lysis
C3H (H-2^k)	H-2^k-virus infected	50
	H-2^k-noninfected	5
	H-2^d-virus infected	<5
	H-2^d-noninfected	<5
Balb/c (H-2^d)	H-2^k-virus infected	12
	H-2^k -noninfected	12
	H-2^d-virus infected	40
	H-2^d-noninfected	<5

Cytotoxicity assays were performed using IEL from reovirus-infected mice that had been cultured for 5 days in vitro with reovirus-pulsed antigen presenting cells and interleukin-2. Targets were reovirus-infected or noninfected L929 cells (H-2^k) or KD2SV cells (H-2^d).

1992). These findings suggest that the α/β TCR$^+$ IEL, which are found in both humans and mice, are important for protecting intestinal crypts from enteric virus infection. There are still no data to suggest that γ/δ TCR$^+$ IEL play a role in immunity to reovirus infection.

Oral reovirus infection in both conventionally reared and germ-free mice induced a significant expansion of α/β TCR$^+$ IEL that express certain TCR variable genes, including $V\beta_{12}$ and $V\beta_{17}$ in vivo (CHEN et al. 1997). In vitro depletion using monoclonal antibodies to the $V\beta$ regions revealed that TCR $V\beta^+_{12}$ and $V\beta^+_{17}$ cells were responsible for approximately 37% and 77%, respectively, of the virus-specific cytotoxicity. In addition, $V\beta^+_{11}$ IEL either lacked cytotoxicity or were immunosuppressive. The authors suggested that enteric reovirus infection leads to the oligoclonal expansion of certain subsets of α/β CD8$^+$ T cells. Whether this response is distinct from the systemic response is not known.

There has been some debate concerning the serotype-restriction of reovirus-specific CTL. FINBERG et al. (1982) demonstrated that reovirus-specific CTL are MHC class I restricted, and that serotype-restricted cytotoxicity is directed predominately against the neutralizing epitope of the $\sigma1$ hemagglutinin protein. Later LONDON et al. (1989a) and PARKER and SEARS (1990) showed that reovirus-specific CTL generated as a result of systemic reovirus infection are serotype cross-reactive. Likewise the CTL found in Peyer's patches and the epithelium were not serotype-restricted and lysed both reovirus T1L- and T3D-infected target cells. The authors proposed that the increased sensitivity and reduction of background lysis in the in vitro CTL assays account for differences observed between their studies and those conducted earlier. Serotype cross-reactivity by reovirus-specific CTL was recently confirmed by studies demonstrating that virus-specific CTL can recognize both the serotype-specific $\sigma1$ hemagglutinin protein and a cross-reactive epitope on $\sigma1$NS (HOFFMAN et al. 1996).

In addition to eliciting cytotoxicity, reovirus has been used to study suppression of delayed-type hypersensitivity (DTH) in mice. Oral and systemic administration of UV-inactivated virus led to specific, serotype-restricted immune suppression of DTH that was mediated by virus specific regulatory cells (Greene and Weiner 1980; Rubin et al. 1981). Experiments using reassortant viruses revealed that the suppressive epitope resided in the $\mu 1$ capsid protein encoded by the M2 gene. Renewed interest in the phenomenon of oral tolerance may provide an opportunity for continuing this work.

Collectively these studies have been integral in our current understanding of the generation and characterization of mucosal immune responses of GALT following enteric virus infection in immunocompetent animals (Fig. 1). Mucosal IgA is probably important in neutralizing and eliminating reovirus from the intestine. In addition, Th cells are important in generating both antibody and CTL responses, most likely through the production of cytokines. However, the role of reovirus-specific CTL in mucosal immunity in adult mice remains unclear. Following oral reovirus infections in β_2-microglobulin-deficient ($\beta_2m^{-/-}$) mice MHC class I-restricted $CD8^+$ T cells appear unnecessary for viral clearance from the small intestine (Barkon et al. 1996; Major and Cuff, 1997). We found that $\beta_2m^{-/-}$ mice made significantly higher virus-specific IgA and IgG responses than control mice following oral reovirus T1L infection, suggesting that $CD8^+$ T cells downregulate virus-specific mucosal and systemic humoral immunity (Major and Cuff, 1997). The mechanisms by which $CD8^+$ T cells regulate mucosal immunity remain to be elucidated but could include cytokine production or direct cytotoxicity.

5 Mucosal Immunity in Neonatal Mice

Reovirus infection in adult animals does not serve as a good model to study the development of protective immunity against viral pathology because adult immunocompetent mice generally do not develop disease following infection. However, oral infection of neonatal mice with all reovirus serotypes can result in mild gastroenteritis (Barthold et al. 1993). Neonates infected systemically with reovirus T3D can also develop lethal meningoencephalitis whereas reovirus T1L infection results in the less lethal infection of ependymal cells (Hrdy et al. 1982). Such observed differences in tropism for neuronal tissues have been attributed to the $\sigma 1$ hemagglutinin (Weiner et al. 1980).

Reovirus infection elicits both serotype-specific and cross-reactive immune responses, and these immune factors are important for protecting neonatal mice from reovirus-induced CNS diseases. Gaulton et al. (1986) demonstrated that immunization of female mice with reovirus anti-idiotype antibodies can protect their offspring from lethal encephalitis caused by reovirus T3D infection, presumably through the induction of idiotype$^+$ antibodies (anti-anti-idiotype). Because females were intraperitoneally infected with reovirus, the virus-specific immune response was systemic, and protection was likely mediated by placentally transferred antibodies. Additional studies showed that neonates can be protected against reovirus infection if they are pretreated with reovirus-immune rabbit serum or murine anti-reovirus monoclonal antibody, including antibody against the σ1 protein (Tyler et al. 1989; Virgin et al. 1988). Furthermore, adoptive transfer of immune spleen cells protected neonatal mice from reovirus infection. $CD4^+$ or $CD8^+$ T cell-depleted reovirus-immune splenocytes failed to completely prevent disease in neonates, indicating that both T-cell populations are partially protective (Virgin and Tyler 1991).

A strain of reovirus type 3, T3 clone 9 (T3c9), initially isolated as murine MT3a, France (Hrdy et al. 1979, 1982), induces encephalitis after oral infection (Keroack and Fields 1986). Because of this characteristic T3c9 has been used extensively in experiments identifying immunological factors that protect individuals from CNS disease resulting from extraintestinal dissemination of enteric viruses (Virgin et al. 1988). Immunization of female mice with reovirus T1L prior to mating conferred protection against T3c9 infection in their offspring, indicating that protection was not serotype-restricted (Cuff et al. 1990). Foster nursing experiments demonstrated that neonates born of reovirus-immune dams and nursed on nonimmune dams, as well as pups born of nonimmune dams and fostered on lactating reovirus-immune females, are protected against oral T3c9 infection. Thus protective immunity was passively transferred to neonates both transplacentally and in milk. Interestingly, virus titers were not reduced in the small intestines of pups from dams that received either an oral or systemic T1L infection, or a systemic T3D infection. Therefore inhibition of reovirus replication in the intestine appears to be dependent on serotype-specific IgA antibodies in milk. Furthermore, IEL or Peyer's patch lymphocytes transferred from reovirus-immune mice protected neonates from lethal T3c9 infection (Cuff et al. 1991). Phenotypic characterization revealed that protective lymphocytes from the IEL population were $CD8^+$, Thy-1^+ T cells whereas lymphocytes from Peyer's patches were $CD8^-$ and Thy-1^- and were thus likely to be B cells.

Collectively these experiments demonstrate that virus-specific IgA and Th cell responses are important in mucosal immunity and protection against reovirus infection in neonatal mice. However, in contrast to studies in adult mice, virus-specific $CD8^+$ T cells control intestinal reovirus infection, suggesting that $CD8^+$ T cells are important in the absence of mature mucosal humoral immunity in neonatal mice.

6 Innate Immunity in Immunodeficient Mice

Severe combined immunodeficient mice lack B and T cells and are useful for studying specific and innate mechanisms of immunity to pathogens (Bosma et al. 1983). Experiments using SCID mice have been integral in understanding whether specific immunity is protective against, or is responsible for, tissue pathogenesis that can occur as a result of enteric virus infection. George et al. (1990) demonstrated that lethal hepatitis develops in SCID mice given reovirus T1L or T3c9 orally within 6 weeks after infection. The hepatic lesions developed within 2–5 weeks after infection and ranged from hepatocyte swelling to nonparenchymal cell infiltration. Later studies revealed that the percentage of Thy-1.2$^+$, AsGM1$^+$ (natural killer) NK cells increases within the nonparenchymal cell population in reovirus-infected mice (TATERKA et al. 1995). These NK cells were cytotoxic for reovirus-infected YAC-1 cells in vitro. It is unlikely that the NK cells were responsible for liver disease since depletion of NK cells with anti-AsGM1 prior to reovirus infection did not change hepatic pathology in SCID mice. Enteric reovirus infection also enhances NK cell activity by IEL in SCID mice. However, the intestinal NK cells also do not appear to control progression of hepatic disease (CUFF et al. 1995 and unpublished data).

7 Reovirus Infection as a Model to Study Development of IgA Responses in GALT

IgA is the predominant antibody isotype that develops following antigenic priming of mucosa-associated lymphoid tissue. However the cellular and molecular mechanisms of this phenomenon are still not clear. WEINSTEIN and CEBRA (1991) addressed the question of whether preferential production of IgA is due to intrinsic differences in the mucosa-associated lymphoid tissue microenvironment or is a consequence of continual activation of gut lymphoid cells by comparing Peyer's patch and peripheral lymph node responses after oral or systemic infections. The investigators used germ-free mice to examine developing mucosal immune responses in the absence of other immune responses to environmental antigens.

Unlike conventional mice, Peyer's patches from germ-free mice do not constitutively contain germinal centers. WEINSTEIN and CEBRA (1991) demonstrated that Peyer's patch germinal centers contained cells that bound high levels of peanut hemagglutinin (PNA^{high}) and were sIgA$^+$. These cells appeared by day 6 and receded by day 35 after oral reovirus infection. Secondary reovirus challenge yielded a less robust germinal center reaction in germ-free Peyer's patches, and most of the B cells were PNA^{low}, sIgA$^+$ cells. It is likely that the presence of virus-specific intestinal IgA inhibited a germinal center reaction after reexposure to reovirus due to antigenic exclusion. Following footpad infection, draining lymph nodes developed similar germinal center reactions; however, the predominant antibody isotype was IgG and not IgA. The predisposition of GALT to generate antigen-specific IgA was concluded to be due to the interaction of antigen, antigen-presenting cells, and immune effector cells within the gut microenvironment and was not a function of constitutive activation in these tissues. In addition, these experiments demonstrated that the constitutive presence of Peyer's patch germinal centers required constant antigenic challenge because the germinal center reaction waned once viral infection was resolved in germ-free mice. Moreover, IgA is unlikely to serve as a means for continued antigen presentation by cotransport of antigen into the host via M cells (WELTZIN et al. 1989). However, it is possible that in the adult animal "naturally occurring" IgA antibodies help deliver to antigens to M cells for priming of the mucosal immune system.

Studies reported by HOOPER et al. (1994) supported the hypothesis that continual antigen exposure in the gut affects the activity of Peyer's patch lymphocytes and immune responses. Peyer's patch $CD4^+$ T lymphocytes spontaneously proliferated in autologous mixed lymphocyte cultures in vitro. Oral reovirus infection of germ-free mice resulted in higher spontaneous proliferation of Peyer's patch cells than in noninfected germ-free mice. However, responses of reovirus infected germ-free mice were significantly lower than spontaneous proliferation observed in conventionally reared, virus-infected mice. These results support the hypothesis that the seemingly spontaneous nature of Peyer's patch proliferation in vitro is due to T cell reactivity against environmental antigens. Thus it is likely that the chronic activation of $CD4^+$ Th cells in GALT plays a role in the constitutive germinal center responses observed in Peyer's patches.

Reovirus has also been used to examine the effects of passively acquired immunity on development of mucosal immune responses in neonatal mice. In GALT, germinal centers and IgA-producing cells develop at approximately 3 weeks of age, which corresponds to the time when mice are weaned (PARROTT and MACDONALD 1990; KRAMER and CEBRA 1995a,b). Studies by KRAMER and CEBRA (1995b) suggested that the presence of maternal IgA in milk secretions, most likely specific for gut natural flora, can inhibit the development of IgA responses in neonates. This hypothesis was further supported by the observation that T1L-infected neonatal mice nursed by reovirus-immune dams made negligible amounts of virus-specific IgA in Peyer's patches compared to neonates nursed by nonimmune dams or SCID dams (KRAMER and CEBRA 1995a). In addition, reovirus infected neonates nursed by reovirus nonimmune dams had enhanced levels of total IgA. Thus KRAMER and CEBRA suggested that there is a "bystander" effect associated with reovirus infection and the triggering of total IgA production. Two possible mechanisms for the enhancement of total IgA following neonatal reovirus infection have been postulated: first, damage to M cells, which has been shown to occur during reovirus infection (BASS et al. 1988), may increase permeability of the epithelium to environmental antigens, and second, reovirus-induced cytokines such as transforming growth factor-β may boost the otherwise quiescent neonatal immune system to become active. Therefore it was suggested that passively transferred maternal IgA in milk, but not transplacental IgG, inhibit virus-specific as well as total IgA responses in neonates. These experiments imply that in order for an oral vaccine to efficiently prime the mucosal immune system of the nursing infant it is necessary to present the antigen in a form that is not reactive with maternal antibodies.

Recently PERIWAL and colleagues (1997) have demonstrated that oral administration of microencapsulated reovirus to neonatal mice resulted in virus-specific IgA in GALT and serum. In addition, the generation of virus-specific IgA occurred even if pups were nursed on immunocompetent dams, indicating that microencapsulation can circumvent the inhibitory effects of maternally transferred antibody. These experiments present an approach to developing oral vaccines that efficiently prime neonatal immunity despite the presence of inhibitory maternal antibodies.

8 Mucosal Immunity to Respiratory Infection

Reovirus infection of the respiratory tract induces various types of pathology ranging from no discernible disease to fatal pneumonia. Therefore reovirus is a good pathogen to study mucosal immunity initiated in the respiratory tract. HOBBS and MASCOLI (1965) demonstrated that all three serotypes of reovirus can infect weanling mice following intranasal inoculation. Experimental respiratory infection has also been used to examine the phenomenon of increased susceptibility to secondary bacterial infection following viral respiratory infection. KLEIN et al. (1967, 1969) examined the consequences of intranasal reovirus infection on bacterial clearance following inoculation with aerosolized *Staphylococcus aureus* in adult mice. They found that mice had approximately three times more bacteria in their lungs 7–10 days after intranasal inoculation with reovirus, implying an immunoregulatory effect of virus infection. These studies support the idea that relatively benign viral infections can alter immune function.

More recently, reovirus infection of the respiratory tract has been used to address questions concerning the function of M cells, viral pathogenesis, and host immune responses in the lung. MORIN and coworkers (1994) initially reported that reovirus T1L entered pulmonary M cells and type I epithelial cells lining epithelial spaces in infected rats. This led to the hypothesis that, as in the gut, M cells serve as a portal of entry in the lung. In a second report MORIN et al. (1996) described the pathology associated with intratracheal infection with reovirus T1L and T3D. Both strains induced acute pneumonia in juvenile rats that was marked by leukocyte infiltration, type I epithelial cell death, and type II epithelial cell hyperplasia. However, reovirus T3D induced a more pronounced infiltrate than did T1L virus. In collaboration with MORIN, FARONE and coworkers (1996) examined the effects of reovirus infection on the production of tumor necrosis factor and C-X-C chemokines in vivo and in vitro. Corresponding to the observation that T3D induced a more robust inflammatory response than T1L, reovirus T3D infection also induced higher levels of mRNA for macrophage inflammatory protein 2 in infected rats. In addition, alveolar macrophages pulsed with reovirus in vitro produced increased mRNA for macrophage inflammatory protein 2, interleukin-1, and tumor necrosis factor-α. These studies are significant because they represent approaches to understanding the mechanisms of lung damage during viral infection as well as the genetic basis for viral infection of the lung.

Respiratory reovirus infections in mice have recently been revisited. BELLUM et al. (1996) confirmed the observations of earlier reports that replicating virus can be found in the lung during the first 7–10 days after respiratory reovirus infection. In addition, mononuclear infiltrates consisting of B cells, T cells, and macrophages were found in the airspaces and interstitial tissues in infected lungs, presenting a pathological picture that is similar to infections with more virulent viruses such as influenza. Flow cytometric analyses and cytotoxicity assays on regional lymphoid tissue during infection revealed an infiltration of predominantly $CD8^+$ T cells in the interstitial and alveolar spaces of infected mice that was coincident with the appearance of reovirus-specific pCTL in lung tissue (THOMPSON et al. 1996). Interestingly, these cells expressed the GCT surface marker recognized by monoclonal antibody CD3.5. A small population of $CD4^+$ T cells were also found to express GCT. It has been suggested that GCT expression is a marker for recently activated pCTL or CTL (LONDON et al. 1990). Because the frequency of virus-specific pCTL as determined by limiting dilution analyses is much lower than the percentage of cells expressing GCT following infection, it will be of interest to understand why some cells that are not functionally antigen-specific still express an activated phenotype by flow cytometry.

These data seem to suggest that reovirus polyclonally activates certain mucosal T-cell populations. This view, while unproved, is supported by the observations that reovirus infection induces polyclonal IgA production in neonates, expands conventional $CD8^+$ T-cell populations other than virus-specific pCTL in the intestinal epithelium, and induces expression of the activation marker GCT on a relatively high percentage of T cells. Alternatively, and perhaps more likely, reovirus infection may alter lymphocyte trafficking, resulting in the migration of cells expressing activated phenotypes to regions where they were not previously found.

9 Summary and Future Directions

Experiments using reovirus as a model pathogen in adult, neonatal, and immunodeficient animals have been useful in determining factors that mediate susceptibility and resistance to viral diseases, generation of specific-immunity, and identification of determinants that can regulate generation of specific mucosal immunity. Because reovirus elicits immune responses after either enteric or respiratory infections, it can be used in studies concerning the relationship between mucosal immune responses at these sites. In addition, reovirus infection provides an opportunity to study molecular and cellular events that regulate the induction and expression of mucosal immune responses, such as M-cell function and cytokine production by mucosal T cells. Reovirus infection can also be used to elucidate mechanisms that are involved in the induction and maintenance of oral tolerance using an antigen that induces both humoral and cellular mucosal immunity. Furthermore, reovirus might prove to be an effective vehicle for delivery of mucosal vaccines. In conclusion, many advancements in our current knowledge of mucosal immune responses have been facilitated by studies of respiratory and enteric reovirus infections, and it is likely that reovirus will continue to be used as a probe to understand the function and regulation of one of our most important defenses, mucosal immunity.

Acknowledgements. We thank John Cebra for providing unpublished data for this review. This work was supported by grants AI-34544 (C.F.C.) and AI-23970 (D.H.R.) from the National Institutes of Health and a Merit Review grant from Veterans Affairs.

References

Barkon ML, Haller BL, Virgin HW (1996) Circulating immunoglobulin G can play a critical role in clearance of intestinal reovirus infection. J Virol 70:1109–1116

Barthold SW, Smith AL, Bhatt PN (1993) Infectivity, disease patterns, and serologic profiles of reovirus serotypes 1, 2, and 3 in infant and weanling mice. Lab Anim Sci 43:425–430

Bass DM, Trier JS, Dambrauskas R, Wolf JL (1988) Reovirus type I infection of small intestinal epithelium in suckling mice and its effect on M cells. Lab Invest 58:226–235

Bellum SC, Hamamdzic D, Thompson AH, Harley RA, London SD, London L (1996) Experimental reovirus serotype 1/strain Lang infection of the lung: a model for the study of the lung in the context of mucosal immunity. Lab Invest 74:221–231

Bosma GC, Custer RP, Bosma MJ (1983) A severe combined immunodeficiency mutation in the mouse. Nature 301:527–530

Chen D, Lee F, Cebra JJ, Rubin DH (1997) Predominant T cell receptor Vβ usage of intraepithelial lymphocytes during the immune response to enteric reovirus infection. J Virol 71:3431–3436

Cuff CF, Lavi E, Cebra CK, Cebra JJ, Rubin DH (1990) Passive immunity to fatal reovirus serotype 3-induced meningoencephalitis mediated by both secretory and transplacental factors in neonatal mice. J Virol 64:1256–1263

Cuff CF, Cebra CK, Lavi E, Molowitz EH, Rubin DH, Cebra JJ (1991) Protection of neonatal mice from fatal reovirus infection by immune serum and gut derived lymphocytes. Adv Exp Med Biol 310:307–315

Cuff CF, Hooper DC, Kramer D, Rubin D, Cebra JJ (1992) Functional and phenotypic analyses of the mucosal immune response in mice: approaches to studying the immunogenicity of antigens applied by the enteric route. Vaccine Res 1:175–182

Cuff CF, Cebra CK, Rubin DH, Cebra JJ (1993) Developmental relationship between cytotoxic alpha/beta T cell receptor-positive intraepithelial lymphocytes and Peyer's patch lymphocytes. Eur J Immunol 23:1333–1339

Cuff CF, Rubin DH, Cebra JJ (1995) Development, function, and trafficking of mucosal leukocytes in immunodeficient mice. Adv Exp Med Biol 371B:719–724

Dimmock NJ (1993) Neutralization of animal viruses. In: Dimmock NJ (ed) Neutralization of animal viruses. Springer, Berlin Heidelberg New York, pp 1–149 (Current topics in microbiology and immunology, vol 183)

Farone AL, Frevert CW, Farone MB, Morin MJ, Fields BN, Paulauskis JD, Kobzik L (1996) Serotype-dependent induction of pulmonary neutrophilia and inflammatory cytokine gene expression by reovirus. J Virol 70:7079–7084

Finberg R, Weiner HL, Fields BN, Benacerraf B, Burakoff SJ (1979) Generation of cytolytic T lymphocytes after reovirus infection: role of S1 gene. Proc Natl Acad Sci USA 76:442–446

Finberg R, Spriggs DR, Fields BN (1982) Host immune response to reovirus: CTL recognize the major neutralization domain of the viral hemagglutinin. J Immunol 129:2235–2238

Gaulton G, Sharpe A, Chang D, Fields BN, Greene MI (1986) Syngeneic monoclonal internal image anti-idiotopes as prophylactic vaccines. J Immunol 37:2930–2936

George A, Kost SI, Witzleben CL, Cebra JJ, Rubin DH (1990) Reovirus-induced liver disease in severe combined immunodeficient (SCID) mice. A model for the study of viral infection, pathogenesis, and clearance. J Exp Med 171:929–934

Greene MI, Weiner HL (1980) Delayed hypersensitivity in mice infected with reovirus. II. Induction of tolerance and suppressor T cells to viral specific gene products. J Immunol 125:283–287

Hayes EC, Lee PWK, Miller SE, Joklik WK (1981) The interaction of a series of hybridoma IgGs with reovirus particles. Demonstration that the core protein lambda 2 is exposed on the particle surface. Virology 108:147–155

Hobbs TR, Mascoli CC (1965) Studies on the experimental infection on weanling mice with reoviruses. Proc Soc Exp Biol Med 118:847–853
Hoffman LM, Hogan KT, Cashdollar LW (1996) The reovirus nonstructural protein sigma1NS is recognized by murine cytotoxic T lymphocytes. J Virol 70:8160–8164
Hooper DC, Rubin DH, Cebra JJ (1994) Spontaneous proliferation of Peyer's patch cells in vitro. Int Immunol 6:873–880
Hrdy DB, Rosen L, Fields BN (1979) Polymorphism of the migration of double-stranded RNA genome segments of reovirus isolates from humans, cattle, and mice. J Virol 31:104–111
Hrdy DB, Rubin DH, Fields BN (1982) Molecular basis of reovirus neurovirulence: role of the M2 gene in avirulence. Proc Natl Acad Sci USA 79:1298–1302
Jackson GG, Muldoon RL, Cooper RS (1961) Reovirus type 1 as an etiologic agent of the common cold. J Clin Invest 40:1051
Keroack M, Fields BN (1986) Viral shedding and transmission between hosts determined by reovirus L2 gene. Science 232:1635–1638
Kilian M, Russell MW (1994) Function of mucosal immunoglobulins. In: Ogra P, Mestecky J, Lamm M, Strober W, McGhee J, Bienenstock J (eds) Handbook of mucosal immunology. Academic Press, San Diego, pp 127–137
Klein JO, Green GM, Tilles JG, Kass EH, Finland M (1967) Bacteriacidal activity of mouse lung in reovirus infection. Antimicrob Agents Chemother 7:164–167
Klein JO, Green GM, Tilles JG, Kass EH, Finland M (1969) Effect of intranasal reovirus infection on antibacterial activity in the mouse lung. J Infect Dis 119:43–50
Kramer DR, Cebra JJ (1995a) Role of maternal antibody in the induction of virus specific and bystander IgA responses in Peyer's patches of suckling mice. Int Immunol 7:911–918
Kramer DR, Cebra JJ (1995b) Early appearance of "natural" mucosal IgA responses and germinal centers in suckling mice developing in the absence of maternal antibodies. J Immunol 154:2051–2062
Leers WD, Royce KR (1966) A survey of reovirus antibodies in sera of urban children. Can Med Assoc J 94:1040–1042
Lerner AM, Cherry JD, Klein JO, Finland M (1947) Infections with reoviruses. N Engl J Med 267:947–952
Letvin NL, Kauffman RS, Finberg R (1981) T lymphocyte immunity to reovirus: cellular requirements for generation and role in clearance of primary infections. J Immunol 127:2334–2339
London SD, Rubin DH, Cebra JJ (1987) Gut mucosal immunization with reovirus serotype 1/L stimulates virus-specific cytotoxic T cell precursors as well as IgA memory cells in Peyer's patches. J Exp Med 165:830–847
London SD, Cebra JJ, Rubin DH (1989a) The reovirus-specific cytotoxic T cell response is not restricted to serotypically unique epitopes associated with the virus hemagglutinin. Microb Pathog 6:43–50
London SD, Cebra JJ, Rubin DH (1989b) Intraepithelial lymphocytes contain virus-specific, MHC-restricted cytotoxic cell precursors after gut mucosal immunization with reovirus serotype 1/Lang. Reg Immunol 2:98–102
London SD, Cebra-Thomas JA, Rubin DH, Cebra JJ (1990) CD8 lymphocyte subpopulations in Peyer's patches induced by reovirus serotype 1 infection. J Immunol 144:3187–3194
Major AS, Cuff CF (1996) Effects of the route of infection on immunoglobulin G subclasses and specificity of the reovirus-specific humoral immune response. J Virol 70:5968–5974
Major AS, Cuff CF (1996) Enhanced mucosal and systemic immune responses to intestinal reovirus infection in β2-microglobulin-deficient mice. J Virol 71:5782–5789
Margolis G, Kilham L, Gonatas NK (1971) Reovirus type III encephalitis: observations of virus-cell interactions in neuronal tissue I. Light microscopy studies. Lab Invest 24:91–100
Mazanec MB, Kaetzel CS, Lamm ME, Fletcher D, Nedrud JG (1992) Intracellular neutralization of virus by immunoglobulin A antibodies. Proc Natl Acad Sci USA 89:6901–6905
Morin MJ, Warner A, Fields BN (1994) A pathway for entry of reoviruses into the host through M cells of the respiratory tract. J Exp Med 180:1523–1527
Morin MJ, Warner A, Fields BN (1996) Reovirus infection in rat lungs as a model to study the pathogenesis of viral pneumonia. J Virol 70:541–548
Onodera T, Toniolo A, Ray UR, Jenson AB, Knazek RA, Notkins AL (1981) Virus-induced diabetes mellitus. XX. Polyendocrinopathy and autoimmunity. J Exp Med 153:1457–1473
Outlaw MC, Dimmock NJ (1990) Mechanisms of neutralization of influenza virus on mouse tracheal epithelial cells by mouse monoclonal polymeric IgA and polyclonal IgM directed against the viral haemagglutinin. J Gen Virol 71:69–76

Outlaw MC, Dimmock NJ (1991) Insights into neutralization of animal viruses gained from study of influenza virus. Epidemiol Infect 106:205–220
Outlaw MC, Armstrong SJ, Dimmock NJ (1990) Mechanisms of neutralization of influenza virus in tracheal epithelial and BHK cells vary according to IgG concentration. Virology 178:478–485
Parker SE, Sears DW (1990) H-2 restriction and serotype crossreactivity of anti-reovirus cytotoxic T lymphocytes (CTL). Viral Immunol 3:77–87
Parrott D, MacDonald TT (1990) The ontogeny of the mucosal immune system in rodents. In: MacDonald TT (ed) Ontogeny of the immune system of the gut. CRC Press, Boca Raton, pp 51–60
Periwal SB, Speaker TJ, Cebra JJ (1997) Orally administered microencapsulated reovirus can bypass suckled neutralizing maternal antibody that inhibits active immunization of neonates. J Virol 71:2844–2850
Rubin D, Weiner HL, Fields BN, Greene MI (1981) Immunologic tolerance after oral administration of reovirus: requirement for two viral gene products for tolerance induction. J Immunol 127:1697–1701
Rubin DH, Anderson AO, Lucis D (1983) Potentiation of the secretory IgA response by oral and enteric administration of CP 20,961. Ann NY Acad Sci 409:866–870
Rubin DH, Kornstein MJ, Anderson AO (1985) Reovirus serotype 1 intestinal infection: a novel replicative cycle with ileal disease. J Virol 53:391–398
Rubin DH, Eaton MA, Anderson AO (1986) Reovirus infection in adult mice: the virus hemagglutinin determines the site of intestinal disease. Microb Pathog 1:79–87
Selb B, Weber B (1994) A study of human reovirus IgG and IgA antibodies by ELISA and Western blot. J Virol Methods 47:15–26
Stanley NF, Joske RA (1975) Animal model of human disease: chronic biliary obstruction caused by reovirus type 3. Am J Pathol 80:185–187
Taterka J, Cebra JJ, Rubin DH (1995) Characterization of cytotoxic cells from reovirus-infected SCID mice: activated cells express natural killer- and lymphokine-activated killer-like activity but fail to clear infection. J Virol 69:3910–3914
Thompson AH, London L, Bellum SC, Hamamdzic D, Harley RA, London SD (1996) Respiratory-mucosal lymphocyte populations induced by reovirus serotype 1 infection. Cell Immunol 169:278–287
Tyler KL, Virgin HW, Bassel-Duby R, Fields BN (1989) Antibody inhibits defined stages in the pathogenesis of reovirus serotype 3 infection of the central nervous system. J Exp Med 170:887–900
Tyler KL, Mann MA, Fields BN, Virgin HW (1993) Protective anti-reovirus monoclonal antibodies and their effects on viral pathogenesis. J Virol 67:3446–3453
Virgin HW, Tyler KL (1991) Role of immune cells in protection against and control of reovirus infection in neonatal mice. J Virol 65:5157–5164
Virgin HW, Bassel-Duby R, Fields BN, Tyler KL (1988) Antibody protects against lethal infection with the neurally spreading reovirus type 3 (Dearing). J Virol 62:4594–4604
Virgin HW, Mann MA, Fields BN, Tyler KL (1991) Monoclonal antibodies to reovirus reveal structure/function relationships between capsid proteins and genetics of susceptibility to antibody action. J Virol 65:6772–6781
Virgin HW, Mann MA, Tyler KL (1994) Protective antibodies inhibit reovirus internalization and uncoating by intracellular proteases. J Virol 68:6719–6729
Weiner HL, Powers ML, Fields BN (1980) Absolute linkage of virulence and central nervous system cell tropism of reoviruses to viral hemagglutinin. J Infect Dis 141:609–616
Weinstein PD, Cebra JJ (1991) The preference for switching to IgA expression by Peyer's patch germinal center B cells is likely due to the intrinsic influence of their microenvironment. J Immunol 147:4126–4135
Weltzin R, Lucia-Jandris P, Michetti P, Fields BN, Kraehenbuhl JP, Neutra MR (1989) Binding and transepithelial transport of immunoglobulins by intestinal M cells: demonstration using monoclonal IgA antibodies against enteric viral proteins. J Cell Biol 108:1673–1685
Wolf JL, Rubin DH, Finberg R, Kauffman RS, Sharpe AH, Trier JS, Fields BN (1981) Intestinal M cells: a pathway for entry of reovirus into the host. Science 212:471–472
Wolf JL, Dambrauskas R, Sharpe AH, Trier JS (1987) Adherence to and penetration of the intestinal epithelium by reovirus type 1 in neonatal mice. Gastroenterology 92:82–91

Subject Index

Subject Index of Companion Volume 233/I

Current Topics in Microbiology and Immunology

Volumes published since 1989 (and still available)

Vol. 194: **Potter, Michael; Melchers, Fritz (Eds.):** Mechanisms in B-cell Neoplasia. 1995. 152 figs. XXV, 458 pp. ISBN 3-540-58447-1

Vol. 195: **Montecucco, Cesare (Ed.):** Clostridial Neurotoxins. 1995. 28 figs. XI., 278 pp. ISBN 3-540-58452-8

Vol. 196: **Koprowski, Hilary; Maeda, Hiroshi (Eds.):** The Role of Nitric Oxide in Physiology and Pathophysiology. 1995. 21 figs. IX, 90 pp. ISBN 3-540-58214-2

Vol. 197: **Meyer, Peter (Ed.):** Gene Silencing in Higher Plants and Related Phenomena in Other Eukaryotes. 1995. 17 figs. IX, 232 pp. ISBN 3-540-58236-3

Vol. 198: **Griffiths, Gillian M.; Tschopp, Jürg (Eds.):** Pathways for Cytolysis. 1995. 45 figs. IX, 224 pp. ISBN 3-540-58725-X

Vol. 199/I: **Doerfler, Walter; Böhm, Petra (Eds.):** The Molecular Repertoire of Adenoviruses I. 1995. 51 figs. XIII, 280 pp. ISBN 3-540-58828-0

Vol. 199/II: **Doerfler, Walter; Böhm, Petra (Eds.):** The Molecular Repertoire of Adenoviruses II. 1995. 36 figs. XIII, 278 pp. ISBN 3-540-58829-9

Vol. 199/III: **Doerfler, Walter; Böhm, Petra (Eds.):** The Molecular Repertoire of Adenoviruses III. 1995. 51 figs. XIII, 310 pp. ISBN 3-540-58987-2

Vol. 200: **Kroemer, Guido; Martinez-A., Carlos (Eds.):** Apoptosis in Immunology. 1995. 14 figs. XI, 242 pp. ISBN 3-540-58756-X

Vol. 201: **Kosco-Vilbois, Marie H. (Ed.):** An Antigen Depository of the Immune System: Follicular Dendritic Cells. 1995. 39 figs. IX, 209 pp. ISBN 3-540-59013-7

Vol. 202: **Oldstone, Michael B. A.; Vitković, Ljubiša (Eds.):** HIV and Dementia. 1995. 40 figs. XIII, 279 pp. ISBN 3-540-59117-6

Vol. 203: **Sarnow, Peter (Ed.):** Cap-Independent Translation. 1995. 31 figs. XI, 183 pp. ISBN 3-540-59121-4

Vol. 204: **Saedler, Heinz; Gierl, Alfons (Eds.):** Transposable Elements. 1995. 42 figs. IX, 234 pp. ISBN 3-540-59342-X

Vol. 205: **Littman, Dan R. (Ed.):** The CD4 Molecule. 1995. 29 figs. XIII, 182 pp. ISBN 3-540-59344-6

Vol. 206: **Chisari, Francis V.; Oldstone, Michael B. A. (Eds.):** Transgenic Models of Human Viral and Immunological Disease. 1995. 53 figs. XI, 345 pp. ISBN 3-540-59341-1

Vol. 207: **Prusiner, Stanley B. (Ed.):** Prions Prions Prions. 1995. 42 figs. VII, 163 pp. ISBN 3-540-59343-8

Vol. 208: **Farnham, Peggy J. (Ed.):** Transcriptional Control of Cell Growth. 1995. 17 figs. IX, 141 pp. ISBN 3-540-60113-9

Vol. 209: **Miller, Virginia L. (Ed.):** Bacterial Invasiveness. 1996. 16 figs. IX, 115 pp. ISBN 3-540-60065-5

Vol. 210: **Potter, Michael; Rose, Noel R. (Eds.):** Immunology of Silicones. 1996. 136 figs. XX, 430 pp. ISBN 3-540-60272-0

Vol. 211: **Wolff, Linda; Perkins, Archibald S. (Eds.):** Molecular Aspects of Myeloid Stem Cell Development. 1996. 98 figs. XIV, 298 pp. ISBN 3-540-60414-6

Vol. 212: **Vainio, Olli; Imhof, Beat A. (Eds.):** Immunology and Developmental Biology of the Chicken. 1996. 43 figs. IX, 281 pp. ISBN 3-540-60585-1

Vol. 213/I: **Günthert, Ursula; Birchmeier, Walter (Eds.):** Attempts to Understand Metastasis Formation I. 1996. 35 figs. XV, 293 pp. ISBN 3-540-60680-7

Vol. 213/II: **Günthert, Ursula; Birchmeier, Walter (Eds.):** Attempts to Understand Metastasis Formation II. 1996. 33 figs. XV, 288 pp. ISBN 3-540-60681-5

Vol. 213/III: **Günthert, Ursula; Schlag, Peter M.; Birchmeier, Walter (Eds.):** Attempts to Understand Metastasis Formation III. 1996. 14 figs. XV, 262 pp. ISBN 3-540-60682-3

Vol. 214: **Kräusslich, Hans-Georg (Ed.):** Morphogenesis and Maturation of Retroviruses. 1996. 34 figs. XI, 344 pp. ISBN 3-540-60928-8

Vol. 215: **Shinnick, Thomas M. (Ed.):** Tuberculosis. 1996. 46 figs. XI, 307 pp. ISBN 3-540-60985-7

Vol. 216: **Rietschel, Ernst Th.; Wagner, Hermann (Eds.):** Pathology of Septic Shock. 1996. 34 figs. X, 321 pp. ISBN 3-540-61026-X

Vol. 217: **Jessberger, Rolf; Lieber, Michael R. (Eds.):** Molecular Analysis of DNA Rearrangements in the Immune System. 1996. 43 figs. IX, 224 pp. ISBN 3-540-61037-5

Vol. 218: **Berns, Kenneth I.; Giraud, Catherine (Eds.):** Adeno-Associated Virus (AAV) Vectors in Gene Therapy. 1996. 38 figs. IX,173 pp. ISBN 3-540-61076-6

Vol. 219: **Gross, Uwe (Ed.):** Toxoplasma gondii. 1996. 31 figs. XI, 274 pp. ISBN 3-540-61300-5

Vol. 220: **Rauscher, Frank J. III; Vogt, Peter K. (Eds.):** Chromosomal Translocations and Oncogenic Transcription Factors. 1997. 28 figs. XI, 166 pp. ISBN 3-540-61402-8

Vol. 221: **Kastan, Michael B. (Ed.):** Genetic Instability and Tumorigenesis. 1997. 12 figs.VII, 180 pp. ISBN 3-540-61518-0

Vol. 222: **Olding, Lars B. (Ed.):** Reproductive Immunology. 1997. 17 figs. XII, 219 pp. ISBN 3-540-61888-0

Vol. 223: **Tracy, S.; Chapman, N. M.; Mahy, B. W. J. (Eds.):** The Coxsackie B Viruses. 1997. 37 figs. VIII, 336 pp. ISBN 3-540-62390-6

Vol. 224: **Potter, Michael; Melchers, Fritz (Eds.):** C-Myc in B-Cell Neoplasia. 1997. 94 figs. XII, 291 pp. ISBN 3-540-62892-4

Vol. 225: **Vogt, Peter K.; Mahan, Michael J. (Eds.):** Bacterial Infection: Close Encounters at the Host Pathogen Interface. 1998. 15 figs. IX, 169 pp. ISBN 3-540-63260-3

Vol. 226: **Koprowski, Hilary; Weiner, David B. (Eds.):** DNA Vaccination/Genetic Vaccination. 1998. 31 figs. XVIII, 198 pp. ISBN 3-540-63392-8

Vol. 227: **Vogt, Peter K.; Reed, Steven I. (Eds.):** Cyclin Dependent Kinase (CDK) Inhibitors. 1998. 15 figs. XII, 169 pp. ISBN 3-540-63429-0

Vol. 228: **Pawson, Anthony I. (Ed.):** Protein Modules in Signal Transduction. 1998. 42 figs. IX, 368 pp. ISBN 3-540-63396-0

Vol. 229: **Kelsoe, Garnett; Flajnik, Martin (Eds.):** Somatic Diversification of Immune Responses. 1998. 38 figs. IX, 221 pp. ISBN 3-540-63608-0

Vol. 230: **Kärre, Klas; Colonna, Marco (Eds.):** Specificity, Function, and Development of NK Cells. 1998. 22 figs. IX, 248 pp. ISBN 3-540-63941-1

Vol. 231: **Holzmann, Bernhard; Wagner, Hermann (Eds.):** Leukocyte Integrins in the Immune System and Malignant Disease. 1998. 40 figs. XIII, 189 pp. ISBN 3-540-63609-9

Vol. 232: **Whitton, J. Lindsay (Ed.):** Antigen Presentation. 1998. 11 figs. IX, 244 pp. ISBN 3-540-63813-X

Vol. 233/I: **Tyler, Kenneth L.; Oldstone, Michael B. A. (Eds.):** Reoviruses I. 1998. 29 figs. XVII, 223 pp. ISBN 3-540-63946-2

MIX
Papier aus verantwortungsvollen Quellen
Paper from responsible sources
FSC® C105338

If you have any concerns about our products,
you can contact us on
ProductSafety@springernature.com

In case Publisher is established outside the EU,
the EU authorized representative is:
Springer Nature Customer Service Center GmbH
Europaplatz 3, 69115 Heidelberg, Germany

Printed by Libri Plureos GmbH
in Hamburg, Germany